D0551210

MADE AT HOME
EGGS & POULTRY

Thank you to everyone who has taken the time to teach us so that we can now pass it on to Indy and anyone else who will listen

Eggs & Poultry
by Dick & James Strawbridge

First published in Great Britain in 2012
by Mitchell Beazley, an imprint of Octopus
Publishing Group Limited, Endeavour House,
189 Shaftesbury Avenue, London, WC2H 8JY
www.octopusbooks.co.uk

An Hachette UK Company
www.hachette.co.uk

ISBN: 978-1-84533-655-4

A CIP catalogue record for this book is available from the British Library

Printed and bound in China

Neither the authors nor the publishers take any responsibility for any injury or damage resulting from the use of techniques shown or described in this book.

Both metric and imperial measurements are given for the recipes. Use one set of measures only, not a mixture of both.

Standard level spoon measurements are used in all recipes
1 tablespoon = 15ml
1 teaspoon = 5ml

Ovens should be preheated to the specified temperature. If using a fan-assisted oven, follow the manufacturer's instructions for adjusting the time and temperature. Grills should also be preheated.

This book includes dishes made with nuts and nut derivatives. It is advisable for those with known allergic reactions to nuts and nut derivatives and those who may be potentially vulnerable to these allergies, such as pregnant and nursing mothers, invalids, the elderly, babies and children to avoid dishes made with nuts and nut oils.

It is also prudent to check the labels of preprepared ingredients for the possible inclusion of nut derivatives.

The Department of Health advises that eggs should not be consumed raw. This book contains some dishes made with raw or lightly cooked eggs. It is prudent for more vulnerable people such as pregnant and nursing mothers, invalids, the elderly, babies and young children to avoid uncooked or lightly cooked dishes made with eggs.

MADE AT HOME

DICK & JAMES STRAWBRIDGE

EGGS & POULTRY

MITCHELL BEAZLEY

CONTENTS

INTRODUCTION

We've always loved keeping poultry. There is no kind way of putting it, but the hens, ducks, geese and turkeys we have kept have not been the brightest of animals, but a lot of them were real characters. Our first hens were 'rescued' from a farm that kept them in poor conditions. That sounds very noble, but it was much more a matter of paying a farmer and taking 3 hens to a better life.

That was nearly 30 years ago, and Flora, Frieda and Felicity produced massive eggs and were completely free-range. Having been kept in a cage with 1 square foot of space each, the girls must have been mystified by our garden, which was a good size with a number of apple trees. The first morning they were given access to it, they didn't move. We watched with bated breath and it was a couple of hours before they started scratching and exploring, but from then on they had a ball. As our decision to bring back the hens was made on the spur of the moment, their new house had to be constructed quickly, using some old packing cases we had in the garage. It was far from salubrious, but it was dry, draught-free and big enough for them to have room to lay and perch. The house cost nothing, but we did have to spend a little bit on chicken wire for a small outside run (we reasoned that a small run was acceptable as the girls spent most of the day scratching around the garden and hedges).

By rights we probably shouldn't have been keeping hens in our garden, as it was rented accommodation. Lots of our neighbours had stripy lawns and neat flower beds, whereas we had a vegetable patch and hens that laid large eggs and deposited equally large droppings. But we subscribe to the old adage that 'where there's a will, there's a way' and have kept hens ever since.

After a cockerel come to stay and a hen went broody, we had our first brood of chicks. As they started to mature, half of them began making rather unimpressive cock-a-doodle-doos. In a neighbourhood where people don't expect you to have anything other than a neat lawn, we were very aware that our flock was becoming noisier. All it took was a very early morning alarm call from our cockerels for us to decide it was time to eat them. Young cockerels are actually very skinny, especially when slowly reared and completely free-range. They tasted great.

WHERE TO START

Reading books on the subject can give you the confidence you need to try raising poultry, but it is only when you start keeping birds that you really find out if you enjoy handling them or not. It's not that you have to spend your days picking them up – we go weeks without actually touching a bird – 'handling' means getting them to do what you want. Our hens usually go out to explore in the morning and then go to bed by themselves, but even now some of them will still end up on the wrong side of the fence or not go to bed properly, so we have to either shepherd them back to their coop or pick them up and pop them in.

The ducks and geese, on the other hand, have never gone to bed of their own volition, and they do have to be locked up at night because of foxes. We don't have that many turkeys – usually a dozen, as they make great Christmas presents – and they need to be herded and calmly encouraged into their shed at night. Our days are punctuated by letting out and feeding the animals in the mornings, and then making sure they are all safely tucked up in the evenings. The routine is easy to establish, although it does take a little longer in the beginning, on those first few nights after the young birds (also known as poults) have arrived. As well as having to clip nails or spurs, you will undoubtedly have to deal with poorly birds at some point, or just have to catch them for their own good (did we mention they are a bit thick and get into pickles?). The more you handle poultry, the more confident you will become, scooping up a bird without a second thought. That said, a large, grumpy gander or a 25-pound turkey can be a bit of a handful.

THE BENEFITS OF KEEPING POULTRY

There are lots of reasons for keeping birds. You may just want to know where your eggs or meat come from, you could be into rare breeds or there may be a particular bird you like the look of and have always wanted to keep – you may even have inherited some birds you felt sorry for. Whatever the reason, there are many ways of keeping and housing poultry. As keeping hens domestically is becoming increasingly popular, there are a selection of designer hen houses on the market and any poultry magazine will have numerous adverts for coops and runs. Apart from when it came to Flora, Frieda and Felicity, building our coops and runs has for us always been part of getting ready for our new arrivals, so rest assured housing them can be very economical.

We get great pleasure from keeping our poultry, and the eggs and meat are of the highest quality. If you are going to use the meat and eggs you produce, it is important to have confidence and do them justice by making really tasty meals. Tasty does not mean overly complicated: the first time you have a boiled goose egg with toast you will marvel at what you have been missing. Creamy scrambled eggs made with the golden yolks from your own hens are a treat for breakfast and will spoil you for anything you eat elsewhere. We are very comfortable butchering our own meat and will try anything – it's a mind-set. Boning a bird, for example, is worth the effort. A boned, stuffed turkey follows exactly the same principles as a chicken – it's just bigger and therefore a lot less fiddly. And let's not forget, if you make a mistake, it will still taste great: if you've left some meat on the carcass you can always scrape it off, add it to the stuffing and say you've done it on purpose. Whatever you do, make sure you enjoy the fact that you have taken control of your food and know exactly where it comes from.

Dick & James

PREPARING TO

KEEP POULTRY

Most people start keeping hens because there is nothing quite like a fresh egg from a happy, well-fed chicken. In addition, this chicken, which has been slowly reared and given lots of TLC, will be delicious. Before deciding to keep hens for eggs, let alone turkeys for meat, you have to make sure you are prepared for the reality. Keeping poultry can be messy and smelly at times, and will bring your relationship with your food into sharper focus – when you see your dinner walking around the garden every day, it's hard to kid yourself that meat comes wrapped in cellophane.

Before you decide what kind of birds you want to keep, make sure you have enough space to keep birds at all. A small coop and run for a couple of hens is a way of keeping your family in eggs but it is not ideal – to us free-range means freedom. We have built some small houses and runs that allow us to separate breeding trios (a cockerel and 2 hens that we want to breed together), but we only segregate them for a relatively short period and let them out to roam around on nice days. It is possible to have healthy, contented birds in a relatively small space, but it takes more work to keep them supplied in fresh greens and give them enough exercise. A small movable ark allows you to give your birds fresh vegetation by moving them around your plot. While this does give you control of the areas the poultry will scratch at, they will turn a small area into bare earth very quickly.

Make sure your space and aspirations are compatible. The space determines the size of coop and run, but if you have too many birds, they will trash the ground cover, no matter how large the space. Small movable coops and runs are particularly suited to hens or bantams, but larger fowl need a larger coop: their houses should be airy, draught-free and dry, with sufficient perches and nesting boxes. You can make a poultry house from almost any type of building materials – wooden planks or plywood, bricks or breezeblocks, even stone, corrugated iron or roofing felt. You may wish to avoid using straw bales, as they are too close to the ideal home for rats and mice, although larger ducks and geese will see off smaller vermin. The materials for building housing and runs can be rather expensive, so shop around. If you are thinking of keeping poultry or expanding your flock, it makes sense to gather materials and plan in advance, as this will allow you to build your coops to a similar design, so your field or yard does not resemble a shanty town.

A mind-boggling number of different types of feeders and water containers are available for your flock. The cheap plastic ones are only good for very light usage, so we prefer to buy good-quality galvanized ones whenever possible – some of our feeders are more than 20 years old and still going strong. To keep costs down, you can sometimes pick them up second-hand at agricultural auctions.

BUYING YOUR BIRDS

Buying your birds is great fun. Breeders are always happy to give you advice about keeping their charges, so when you decide the time is right and you are close to choosing what you want to keep, it makes sense to visit a breeder without the intention of bringing anything home. Be strong, as it is all too easy to set off with the intention of buying a hen that is at 'point-of-lay', perfect for providing fresh eggs for your family, only to return with a very pretty bantam that will lay fewer and smaller eggs, just because none of the breeds you had in mind were available and the young bantams caught your eye.

A visit to a poultry auction is also worthwhile: you will get to see a great variety of birds and be able to confirm the size your chosen birds can grow to, as well as having the chance to chat with some experts. While you can buy fertilized eggs for incubation on the internet, buying them at auction from a local supplier can be a better bet, as you can talk to the person you are buying from.

KEEPING DUCKS & GEESE

All poultry need water, but ducks and geese need plenty for bathing, too. We had not kept ducks or geese prior to moving to our smallholding, as we did not have the suitable environment for them. Rather than assuming that you will have to fill water containers every day, it is worth thinking about how water can be diverted to help you and reduce your workload. We widened our very small stream and dug out a rather sizeable pond. The stream stops the pond from becoming stagnant by ensuring there is enough movement to keep it 'flushed' – otherwise, all the goose and duck droppings will just settle to the bottom, and it doesn't take long for the pond to become smelly.

It is very important to take water flow and the location of ponds into account when keeping ducks and geese, but ponds don't necessarily have to be near streams: one of our ponds is supplied with water via a homemade hydraulic ram pump and a good length of hosepipe. The ram pump provides a limited flow, but it pumps all day and night so the pond is kept full. It is half-way up a hill, which allows any overflow to drain away.

BREEDING POULTRY

You may think that all you need is a cockerel and a couple of hens and then let nature take its course. But as always, it is best to be prepared: using a broody hen is probably the easiest way to successfully hatch eggs. A broody hen is a hen whose biological clock has ticked to the point that she reckons she's laid enough eggs and is ready to sit on them and keep them warm until they hatch – the breast of a broody hen gets warmer to help with the job. Some breeds that have been specifically developed for egg laying have had the tendency to go broody bred out of them, but mongrel hens or bantams usually make good broody hens. It will take you a season to know which of your hens are good at being broody. One sign is that she will occupy a nesting box and refuse to move – she may even peck you if you try and move her.

If you have a cockerel of the right breed and good hens that are not related to

him, all you have to do is collect the fertilized eggs and put them under a broody hen. A cockerel will ensure that the eggs of his hens are fertile as that's his raison d'être, so the trick is to collect the eggs of the hens you wish to have bred to the cockerel – this sounds easy, but hens do have a tendency to lay together, so you may have to design your layout to allow segregation.

If you are serious about hatching your own chicks you may consider getting an incubator. Our first incubator cost us next to nothing – we found it in the classified ads of the local paper. It was a very simple box with a heater and a thermostat, and we were responsible for rotating the eggs and keeping the moisture levels correct. It took constant attention, but we were relatively successful. We have since upgraded to a model that is much easier to clean. Thanks in part to the heat and the moisture, hatching eggs is a smelly, messy business – especially once the chicks start emerging.

We have successfully hatched our own chicken, duck and geese eggs, but we have never tried hatching turkey eggs. Instead, we leave others to get our turkeys through the early stage of their lives, preferring to buy young birds or poults at 5 or 6 weeks of age, when they are off heat – that is, big enough to no longer require a heat lamp to keep them warm. And we do have a slight problem in that we always eat our turkeys before they are ready to start breeding.

PREPARING POULTRY

We recommend that you try all the sections on preparing and cooking poultry prior to keeping your first birds. We have been doing our own butchery for years and it is only practice that will give you confidence.

It may seem like stating the obvious, but you really do need to own some good, sharp knives. You don't need many knives to do all the things you will ever want to: a boning knife, a cook's knife and a paring knife (add to that a filleting knife and a butcher's knife if you want to fill your block).

Make sure that your knives are always well sharpened. If you are boning, you will need to keep sharpening the blade as you go, as contact with the bone will take the edge off your knife. There are lots of good-quality knives on the market, it's all down to personal choice.

CHICKENS

INTRODUCTION TO

CHICKENS

Why keep chickens? People all around the world have been keeping hens for thousands of years. The Greeks, Egyptians, Chinese and Romans all enjoyed the perks of rearing their own chickens for eggs and meat. Nowadays they are still great animals to look after, regardless of whether you live in an urban or a rural area. We have been keeping chickens for years, in all sorts of different places. In small back gardens of rented houses, we would arrive and knock up a quick chicken coop for our collection of hens. However you like your eggs in the morning, you can be sure that if they come from your own free-range chickens nothing will compare to the great fresh taste!

PROS

- Easy animal to look after, with very few health problems. Delicious eggs and meat!
- Free chicken manure, which is a great activator for the compost bins.
- Great in early spring for eating grubs and pests inside a greenhouse or around the vegetable plot.

CONS

- Responsibility means that you have to be at home in the morning to let them out and at night to put them away.
- Cost of feed.
- Stepping in chicken poo, if it is a small garden that you share with them.

BASICS

The first decision to make before you go out and buy some chickens is what type you want and how many you are planning to keep. We generally try to mix our small flocks with pure breeds and some super egg-laying hybrids. Hybrids are great layers that are used in commercial farms, and you can expect a large number of eggs from them, while more traditional breeds look more interesting and are useful to have as broody hens – i.e. they make good mothers. We tend to opt for bantams, which are smallish breeds, such as Light Sussex and Rhode Island Red. They lay smaller eggs and are therefore ignored by most commercial farmers. However, we find they are relatively low-cost to feed and you don't need much space to keep them. If you only have a small space, we would recommend getting a few hens and don't waste space on a cockerel.

GUIDELINES

- Don't keep more hens than you need or overcrowd their coop.
- Don't place perches directly below one another.
- Don't feed them any scraps that have been contaminated in the kitchen.
- Keep a fresh supply of water for your hens.
- Protect them from predators with fencing.
- Lock them away at night.
- Provide them with a dust bath.

Chickens need food, grit and water

- Clean them out regularly.
- Clip a cockerel's spurs with nail clippers if they get too long.

ESSENTIAL EQUIPMENT

- Chicken wire or electric fencing
- Galvanized feeder
- Water container
- Straw
- Sawdust
- Grit box or oyster shell
- Coop
- Perches
- Nest box
- Dust bath

CHOOSING WHAT TO BUY

- Rhode Island Reds are good layers as well as being a decent size for the table.
- Light Sussex are a lovely old English breed and another great dual-purpose bird.
- Cuckoo Marans are one of our favourites. They tend to be very hardy and lay lovely large, deep brown eggs. You will need good fencing, though, as they have an adventurous streak and love to escape.
- Warrens are a reliable hybrid and lay well.
- Black Rocks are another very useful hybrid to keep.
- Buff Orpingtons are placid birds that lay smallish eggs. They're also excellent hens for fostering chicks, but if you don't want a broody hen, avoid this breed.

DAY-OLDS

Choosing what age the chickens you get should be is an important option. Day-old chicks are very cheap to buy but are unsexed, so you won't find out which ones are going to grow into cockerels until later. This is a good way to start keeping chickens if you want some birds for meat, but bad news if you simply want plenty of eggs. There is also the additional expense of providing heat for them when they're young and extra feed until they reach point-of-lay.

PULLETS

Pullets are about 8–20 weeks old and it's easy to get a batch of just females. They will take a little more feeding before they start producing eggs, but are a good age to buy.

POINT-OF-LAY PULLETS

At 20 weeks, hens start laying eggs and are at the stage that we call point-of-lay. We have found that buying some point-of-lay birds is well worthwhile if your existing hens are getting a bit old and you want a few more daily eggs! Their first season tends to be their most productive, but you have to spend a bit more money to get point-of-lay pullets. You also have to be patient when they start laying – to begin with the eggs can be very thinly shelled, oddly shaped and a bit irregular. Provide your hens with grit if you want to help them develop strong eggs.

FOOD

Traditionally, chickens were found running around the backyard near the kitchen door and would have been fed on little scraps and spare bits of grain. We still give them some raw vegetable kitchen waste, being extremely careful to keep it away from any other foods to avoid contamination. Supplement their free-range diet of seeds, insects and bugs with a combination of mixed corn and layers pellets. Layers pellets are a more expensive feed supply that help the chickens to lay more eggs. They are fed through a suspended tube feeder that is kept indoors away from droppings and hanging above ground level to make it rat-proof. The pellets feed through

slowly so that you can fill the feeder up at the beginning of the week and then add more when the chickens have finished. Also give your chickens a small pot of grit and oyster shell, to aid with digestion in their gizzard and to avoid calcium deficiency – or thin egg shells.

WATER

It is extremely important that chickens have clean water. Having an automatic watering system plumbed in saves a huge amount of time. But if you fill their water up manually, make sure this is done daily.

FREE-RANGE

Try to provide your chickens with as much free roaming space as possible. In a small garden this involves keeping hens in a movable ark with a covered run. In a suburban setting you should start to open their run in the morning, giving them access to a larger part of the garden, and then lock them away at night. Perhaps the best scenario is to give them a substantial area in which to be free-range (see pages 24–25).

KEEPING A COCKEREL

A cockerel is a great way of keeping your flock together if they are in a free-range area. It will shepherd them and protect them from predators. A cockerel will also give you the opportunity to rear more chicks if you let a hen go broody. The obvious downside is that cockerels can be noisy. If you want to discourage a cockerel from being too loud at night-time, make their coop so that it has a lower roof above his perch. This will prevent him from being able to lift up his neck and crow. Also consider how much your neighbourhood will like you if you rear a cockerel in close quarters – we have been there and done that, and offer apologies to all our previous next-door neighbours.

METHOD #1

CHICKEN COOPS & RUNS

Keeping chickens does not require a great deal of space. They need a coop in which to roost at night, both for shelter and security, and an outdoor space during the day. Some coops are designed with an integral run, while others have a separate structure of wire netting on a wooden framework, which is butted up against the house to create the same effect.

PLANNING A COOP

When designing and building your own coop, the main objective is to provide a comfortable, draught-free, well-ventilated and predator-proof place in which the birds can live when they go indoors.

The first thing to determine is the size of the coop you need, and it is always best to err on the side of being generous and give plenty of floor space. Overcrowding must be avoided, as it will affect the health of the birds and their productivity. If birds have access to an outdoor run, a good standard is 0.186 square metres (2 square feet) per bird.

Poultry other than ducks and geese prefer to roost at night. The perches should be removable so that they can be burnt and replaced if there is an attack of red mite. The perches should be 5cm (2 inches) in diameter and not placed where droppings are an issue.

You will need a minimum of 2 nest boxes, or at least 1 nest box for every 3 birds. Each box should be about 30cm (12 inches) square. Individual nest boxes are better than communal, and there should be a 7.5cm (3 inch) lip at the front. If your set-up is big enough to warrant it, design an access to the nest boxes from the outside, as this will make egg collecting easier.

A red comb and bright eyes indicate that this hen is in lay

The entrance the chickens use to get in and out of the coop is normally called a pop-hole. We use one of two types: a conventional hinged door, and a wooden sliding door that is raised in a runner. The vertical sliding door

A SIMPLE DESIGN FOR A COOP

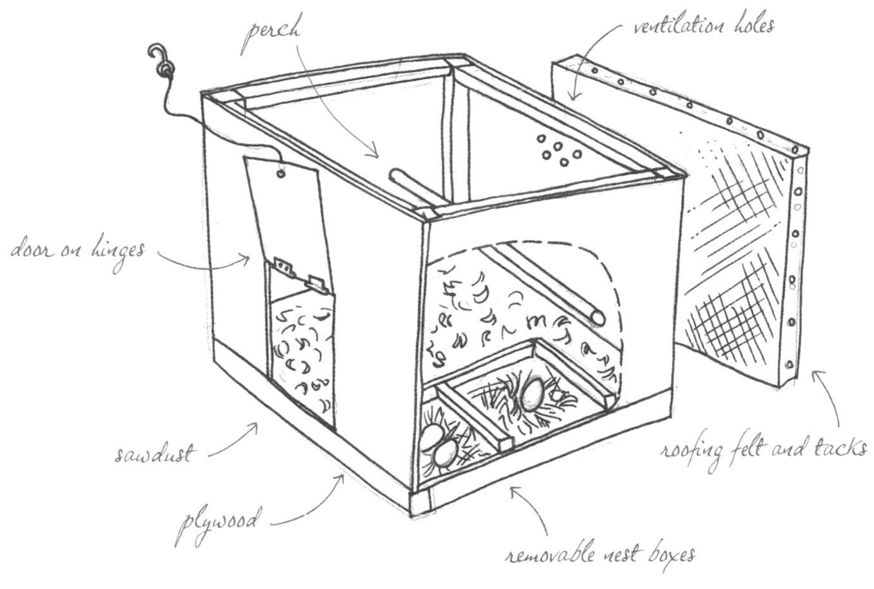

Mobile coops must be moved regularly

is held in place by gravity, which is sufficient to stop predators, such as foxes, prying open the doors.

POSITIONING A COOP

Laying hens are not frightfully noisy; however, if you have a cockerel that is another matter. So it's worth considering your neighbours when choosing a site for your coop or run. It should be dry, but if it is damp, at the very least the coop should be raised off the ground, which has the advantage of not giving rats anywhere to hide. If you intend to have eggs all year you may consider a site near an electricity supply, to provide lighting to extend the winter days. Try to avoid the prevailing wind blowing straight into the building.

Make sure that the position of your coop is convenient. Hens must be let out in the morning and locked away every evening, and because of that your routine will involve regular visits to that area. If your route takes you past vegetable patches, compost heaps or other areas that need to be monitored, the necessity of visiting your hens will have the added advantages of ensuring you are aware of what is happening elsewhere on your plot.

BUILDING A COOP

A poultry house is not difficult to build – all you need is a saw, hammer and nails/staples, screwdriver and screws, and some materials: wood, hinges and chicken wire. Before starting any build it is worth preparing a plan in advance; depending on the materials you have to hand, you can go for a completely new build or you can use those things you have lying around. Your chickens will not complain if their home looks like something from a shanty town, provided you follow some simple principles.

The inside of the coop should be smooth, with as few crevices as possible, as these can provide hiding places for mites that can affect your flock's health. Consider using silicon sealant, or lead-free paint, to seal the joints. The structure of the coop should be draught-free, but if it is a large shed-like coop it should have a window that can be opened to provide ventilation. The design of your roof is up to individual preference – you just need it to slope and be capable of supporting snow if necessary.

PLANNING A RUN

At first glance you may think the best sort of range for poultry is unrestricted access to the outdoors. There are, however, some issues, not least of all the danger from predators.

this willow will grow into a great place for hens

Another disadvantage to giving your chickens the free-range lifestyle is that they will always end up laying eggs in hidden places instead of in their nest boxes. This can be solved if it becomes a problem by keeping them in their run till midday, until they start laying inside again, and letting them out in the afternoon. You can also buy a decoy egg that will encourage them to lay a clutch of fresh ones next to it. Another problem with free-range can be foxes, so you may need to invest in some electric fencing.

A scratching chicken can make short work of a seed bed, or a bed of healthy young plants; indeed, they have a tendency to head straight for the cultivated areas of your plot.

Although access to fresh vegetation and the ability to scratch around for insects is essential for healthy hens, you need to restrict your poultry's movement, and how you do that is determined by the amount of space you have.

Chickens are at home in woodland, and we keep our hens in a willow coppice that provides shade in summer and protection from the elements in winter. They seem very happy there, and the whole area is protected by electric poultry fencing. (We have heard that you can keep foxes at bay by peeing around the boundaries. The best urine apparently comes from meat-eating men ...)

ENCLOSING YOUR RUN

To keep your run protected using wire netting you have two choices: you must either make the fence high enough (and low enough) or have a mesh roof over the run. To build a fox-proof fence you will need to start by digging a trench around the perimeter at least 15cm (6 inches) deep and wide. The fencing needs to be 2m (6 feet) high, and it is better if on the top section a 45cm (18 inch) section is angled outwards at 45 degrees. Netting needs to be laid on the bottom of the trench and then brought upwards. The trench can then be filled and it will stop predators burrowing under the fence.

Electric poultry fencing is meant to protect the hens as much as keep them in. You will need a leisure battery (one that can be charged and discharged lots of times – a car battery will not do), an electric fence unit big enough for the length of fence being used, and poultry fencing, which has smaller mesh at the bottom with larger holes at the top. It is best to cut the grass along the line where you wish to put the fence.

Fences keep hens in and predators out

METHOD #2

PESTS & PROBLEMS

Chickens are good animals to rear because in general they are problem-free. A chicken is hardy and will survive outside in most weather conditions. There are issues that can occur if their house is not kept clean, their diet isn't varied enough or if they are not protected with proper fencing, and all these problems can be quickly dealt with, but the key thing is that prevention is better than cure. We have always tried to clean out the chicken coop as part of a weekly routine, keep fences in good repair and feed our chickens well, primarily so that the only risk to them is when we want to eat them.

RED MITE

Red mite is a small blood-sucking parasite about 0.5mm in size. They hide in small gaps and crevices in chicken coops and feed on roosting birds. We find the best way to deal with them is to keep the coop clean, apply special red mite powder and use plywood instead of planks, to avoid gaps in the coop – you can run but you can't hide!

SCALY FEET

Scaly feet are caused by tiny mites that burrow under the scales on a chicken's legs and through the skin. Lameness can be the severe result, and the best treatment is to isolate affected birds and apply special medication spray from the vet or your local agricultural merchant. Clean the legs afterwards with soapy water and an old toothbrush. In its mildest form, scaly feet can cause discomfort for the hen, and it is far easier to deal with the problem early, before it gets really bad. Another preventative method is to make sure that the hens' perches are clean and their run is washed with disinfectant – air the coop thoroughly before laying down fresh sawdust.

PREDATORS

It used to be the case that foxes were only a problem for chickens in the countryside. However, with the shrinking green spaces, foxes are finding that there is lots of other food in towns and cities, too. Unfortunately, they don't just scavenge for rubbish but will come into your garden and kill your chickens. Other predators that will kill a hen include weasels and sometimes rats.

Preventing the loss of hens to foxes is best countered by good electric fencing, chicken wire, keeping a dog or having a cockerel to protect them. You could set snares, in strict accordance with legal guidelines, but it is vital that you check them daily. Never leave a snare out if you cannot check it, as this is horribly cruel to the animal. Shooting foxes is another option, but this may require a bigger rifle or a shotgun. If using a shotgun, make sure you use cartridges with larger shot. You may find that it is necessary to clip your birds' wings to stop them escaping and becoming vulnerable to predators.

SPUR DAMAGE

Trimming the spurs of your cockerel is very important, in order to protect your hens from getting too badly damaged during mating. It is also pretty important if you have any small free-range children running around. Hold the cockerel firmly – you can wear leather gloves and secure him in a towel if you are worried about him struggling – and use wire cutters to cut back his spurs a little at a time. It is vital that you cut off a small bit at first and gradually work up the spur. Try to avoid opening up his veins, which run in the centre of the spur, and stop at the first sight of blood below the surface. Then file the spurs smooth. If your hens are getting damaged by more than one cockerel and trimming the spurs doesn't remedy the situation, either separate or eat the young males.

SOFT SHELLS

Soft shells are a sign of calcium deficiency. If you notice this when you are collecting your eggs, make sure your birds have a ready supply of oyster shell and grit to increase the strength of the shell. Soft shells are unlikely to be a problem if your chickens are free-range and have access to land outside.

EGG EATING

If a hen starts to eat eggs you need to deal with the problem immediately. Other hens will follow suit unless you separate her from them. The sad reality is that if she returns to the nest boxes after some time apart and still eats eggs, you may need to slaughter and eat her. The only other advice is to provide extra grit in case she is deficient in calcium.

3 hens means 3 eggs most days

METHOD #3

LAYING & BREEDING

The reason we first became interested in keeping our own chickens was for their eggs. The number of hens that we keep fluctuates, depending on the size of our plot and the seasonal variation of the flock, but the aim is always to have enough of our own eggs for a homemade omelette. Providing your hens with the perfect spot to sit and lay is key. You want them to be comfortable, relaxed and to get into a routine. The environment is very similar if you want to breed your hens. You can even hatch chicks without a cockerel by encouraging a broody hen to foster fertilized eggs.

ESSENTIAL EQUIPMENT

- Nest boxes
- Water container
- Feeder
- Grit
- Rat-proof run
- Separate hen house for breeding

EGG LAYING

Going out to the chicken coop and collecting a freshly laid, still warm egg from a nest box is a great way to start any day. It is important to check the boxes daily as eggs that are left for a long time in a nest box are hard to date, may crack and can attract rats or encourage a hen to go broody.

We build our nest boxes to be accessible from outside the coop, so that it's easy to collect the eggs without disturbing all the hens. Chickens like a warm, cosy place to lay eggs, with plenty of fresh straw and the different boxes separated with raised blocks of wood – this also stops the eggs rolling around too much. Try to provide about 1 nest box to every 3 hens, and change the bedding regularly so that the eggs don't absorb any disease through the shell. If your hens seem reluctant to lay, try positioning a decoy egg in the boxes.

Don't expect as many eggs in the winter. Chickens will generally lay every other day, but at colder times of the year this decreases dramatically. If you want more eggs you could light the coop, but we go with the flow and just eat fewer eggs in the winter – after all, chickens aren't machines! Expect the bulk of your eggs from mid spring until late autumn.

You may notice some hens eating their own eggs. This can be a habit that is hard to break, so separate any egg-eaters immediately. Collecting eggs more often and darkening nest boxes can discourage egg eating.

BREEDING

If you want to increase your flock then it really is a matter of letting nature run its course. You can breed chickens any time from late spring to late autumn. A cockerel will mate with any hens in the group, but obviously only similar breeds will produce more of the same type. The mixed chicks from two different breeds will be hardy, and can even in some cases be better egg layers or more suitable table birds. We try to have only one adult cockerel at a time so that we can keep the numbers in check.

If you have a hen that either disappears into a hedge for a few days or occupies a nest box then you've probably got chicks on the way.

BROODY HENS

It's quite easy to tell when one of your hens has gone broody. The key signs to look out for are a hen that will not leave the nest box and fluffs up to twice her size when you peer in. She will make loud noises, and if you pick her up and move her she'll return straight to the eggs she was sitting on. If you want her to hatch chicks, it's important to move her to her own house with the eggs; first, so that the chicks are safe when they hatch, and second, so that she doesn't discourage all the other chickens from coming in to lay eggs for you. Her own house should be warm and cosy, with plenty of straw, plus food and water nearby. We find it's easiest to move broody hens at night.

If you don't want your broody hen to hatch more chicks, you will need to remove the eggs and possibly place her in a box with a slatted base. Raise this up so that the cold air can circulate around her bottom and stop her broody urges.

On average, if left with a hen, eggs will hatch after 21 days. There is very little you need to do, other than provide chick crumb and water for the freshly hatched chicks. Chicks that are hatched naturally tend to be strong and healthy and it is a very easy process, as nature does all the hard work. You can reintroduce the chicks and mother hen to the rest of the flock after 6–8 weeks. Young chicks are vulnerable to rat attacks, so make sure the enclosure is secure.

FOSTER CHICKS

It is possible to put fertilized eggs from another chicken under a broody hen and she will raise them as a foster mother. This doesn't always work, but we have found that some breeds like Buff Orpingtons are particularly good mothers and will accept foster chicks.

METHOD #4

INCUBATING & HATCHING

Incubating eggs is a skill that provides you with a great opportunity to breed your own flock of hens from fertilized eggs. The basic equipment can be bought for a reasonable price. That said, if you are serious about it as a business rather than just a hobby, the quality of your kit will determine your success rates. We have found that the advantages of incubating eggs include the fact that you don't risk a hen going broody in a hedge and getting caught by a fox, and that your hens are not disturbed by the process so you will continue to get fresh eggs to eat. Another massive plus is that chicks are cute – everyone smiles at the little fluff-balls.

ESSENTIAL EQUIPMENT

- Thermometer
- Spray bottle
- Incubator
- Torch
- Chick crumb
- Incubating bulb – infra-red
- Chicken wire brooder

FERTILIZED EGGS

Which comes first: the chicken or the egg? Well, we have always gone down the route of starting with some chickens and increasing our flocks from their fertilized eggs. However, this approach involves keeping a cockerel – difficult if your neighbours are sensitive to early wake-up calls. If you want to start the process from scratch, you can easily buy fertilized eggs online or at a poultry auction and incubate them yourself. This process is slightly more difficult than leaving the hens to rear their own chicks naturally, but it can be very insightful and highly rewarding.

INCUBATOR

We would recommend investing in an electronically controlled incubator to reduce contact time with the eggs. Write the date on the eggs, then introduce them into an already warm incubator. The incubation time is about 21 days, and the ideal temperature in a still air machine is 39.4°C (103°F) (measured with a thermometer 5cm/2 inches above the eggs). It is also necessary to turn the eggs 3 times a day – this enables even heat distribution. The humidity should be about 75–80%.

CANDLING

After 7 days in the incubator you will need to candle the eggs. This means cutting holes in

CANDLING AN EGG

torch

cardboard box with holes

hold the egg inside the box

2 sides of a cardboard box and shining a torch from 1 side through to the other, where you hold an egg. If it is infertile you will see a clear yolk illuminated. If there is a red spot with veins radiating outwards, you have a fertile embryo. Make sure you return the egg to the incubator and wait until it hatches. On the 19th day in the machine, stop turning the egg.

HATCHING

After hatching, leave the chicks by themselves in the incubator for the first 24 hours and don't worry about feeding them. Try to resist taking them out and risking them getting cold. After this drying time, remove the new chicks to a brooder.

BROODER

A brooder is basically a heated nursery where the chicks will live for the next 6 weeks or so. Design your brooder so that there are no sharp corners. If you have several chicks, they could be overcrowded and suffocate each other in the corners. Spread sawdust over the base of the brooder. When you move the chicks across for the first time, show them where the water is by dipping their beaks into it and drop food in front of them, to simulate a mother hen's behaviour.

The brooder must be a rat-proof box with clean wood shavings on the floor and an infra-red lamp suspended above the chicks. Ideally the temperature should be at 35°C (95°F) – reduce this by 3°C (6°F) each week. If it is too hot for the chicks you will see them collecting around the edge of the brooder, away from the lamp. If it is too cold they will be huddled together in the middle. After 6 weeks the chicks should need no more heat. Feed your chicks with chick crumb for the first 6 weeks, then move on to corn and layers pellets.

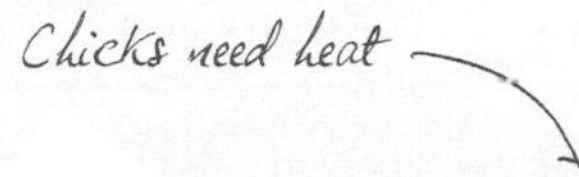

be calm throughout

METHOD #5

SLAUGHTER

If you keep poultry it's a fair bet that you realize meat doesn't come on a plastic tray covered in cellophane. Keeping your own hens does not mean you will ever have to kill any of them; however, if you keep a cockerel and rear your own birds, you will undoubtedly find you end up with a surplus of cockerels that cannot be kept, otherwise the hens will suffer. You could keep them as pets or, like lots of people, try to rehouse them, but there is another option and you will probably never find tastier chicken.

PREPARATION

Before you decide whether or not to kill a bird, you need to make sure that you are prepared and confident that you can do what is necessary. The procedure must be carried out calmly and efficiently, to minimize any stress to the bird.

Killing is best done first thing in the morning, and all food should be taken away the evening before so that the bird's digestive system is not full. Have to hand what you need for plucking and drawing as well, so that once you start you can continue to completion (checklist: sharp knife, bowl, kitchen roll, rubber band, loop of rope at a convenient height to hang bird for plucking, container for feathers, pliers, blowtorch, chopping board, damp cloth). As with all handling of poultry you should be gentle but firm up to the point of actually killing the bird.

Even if you do not intend to keep table birds, if you keep poultry it is sometimes necessary to slaughter an injured bird. This doesn't mean a poorly one, which should be separated into a 'hospital' cage; however, birds can sometimes be injured and may need to be put down.

Small spur buds indicate a young cockerel

HOW TO SLAUGHTER A CHICKEN

It is essential to hold the head correctly, with the neck bones between the first and second fingers and the skull in the palm. Then transfer the hand cradling the bird to its feet.

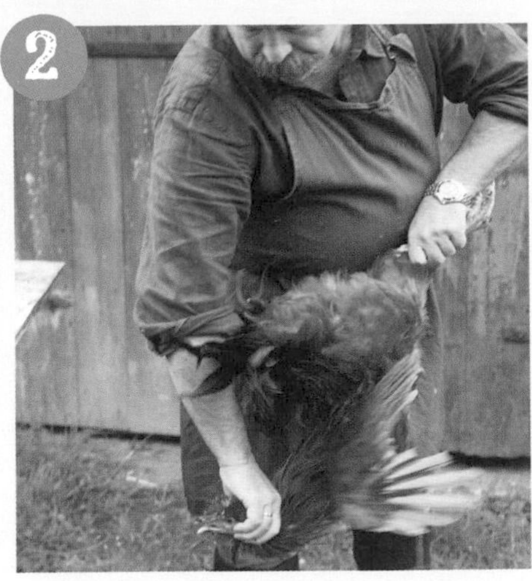

Allow it to hang briefly upside down. Then push down with the hand holding the neck, bending the head back, until the vertebrae separate: the neck will stretch before it breaks.

The bird is dead, but while it is still flapping and the heart is pumping, cut its throat to allow it to bleed out over a bowl.

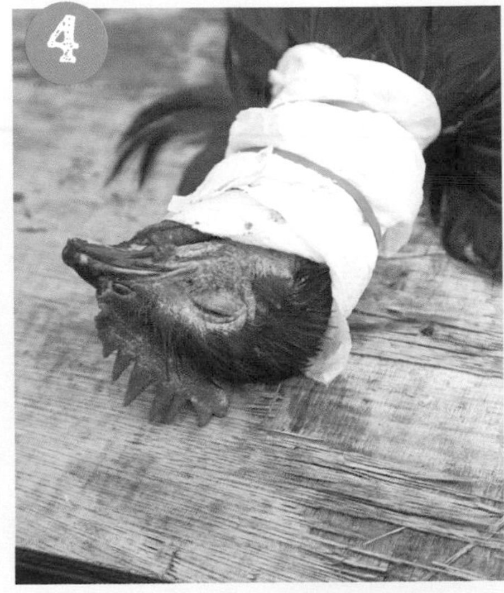

Wrap the throat with kitchen roll and use rubber bands to hold it in place. This will make plucking less messy.

METHOD #6

PLUCKING

The right height is easiest

Plucking a chicken or any poultry is best done when the bird is still warm, otherwise you can tear the skin, and that makes the finished bird look less appealing. There are lots of ways to pluck poultry, some involving submersion in very hot water to loosen the feathers, and others using machines that flail the feathers off the birds, a bit like a mutant car wash. However, we dry pluck, as we believe it to be the most hygienic method and the best suited for small-scale preparation. It has the advantage of requiring next to no resources, and anyone can do it.

HOW TO PLUCK A CHICKEN

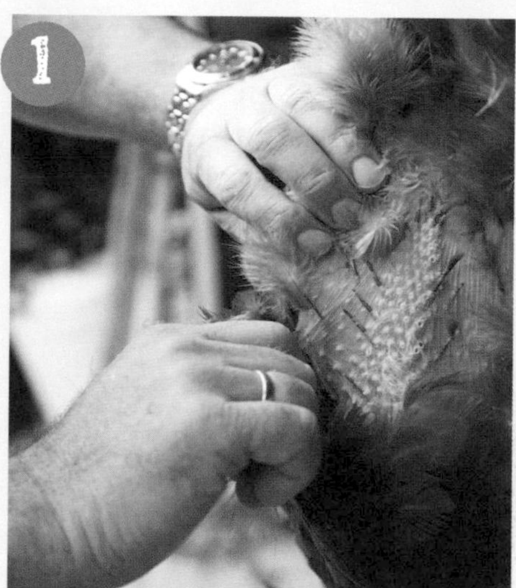

1

With the bird hanging at a convenient height, grab a small tuft of feathers from the breast and give it a short sharp tug downwards.

2

Work your way around the body, continuing to pull downwards and removing the feathers a tuft at a time. With a little practice you will be able to use both hands.

Feathers from the wings can be tough to remove, so take them out one at a time. On a mature bird it may be necessary to use pliers.

It is easiest to pluck the inside of the legs and around the vent if you release one of the legs from the rope.

Singe the fine hair-like feathers off with a blowtorch or a lit piece of rolled-up newspaper. Move the flame quickly so that the flesh does not heat up.

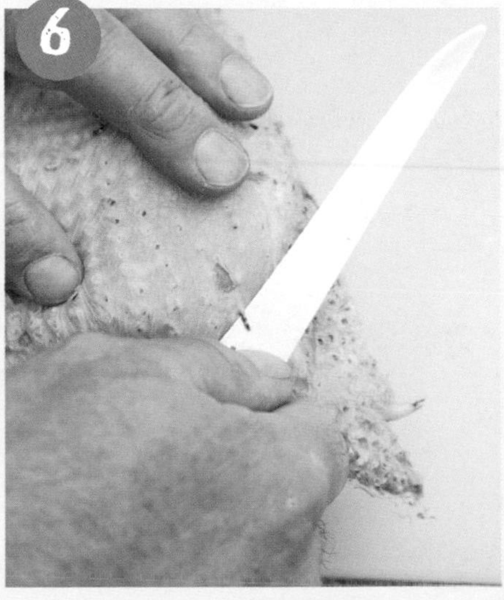

Finally, tidy up the carcass by removing any small quills. The easiest way is to slip a knife blade under them and squeeze between blade and your thumb before yanking them out.

METHOD #7

DRAWING

Removing a bird's entrails is known as drawing. This is not a very pleasant job, but it is necessary, so just get on with it! We do not 'hang' chickens to mature the way we do with larger birds like turkeys, instead we tend to draw them when they are cool – which is less unpleasant than putting your hand inside a warm carcass. When the task is done and the insides are drawn, wipe the carcass with a damp cloth (don't wash it) and save the feet, neck, liver, heart and gizzard, as they can all be eaten or used for stock.

All worthwhile

HOW TO DRAW A CHICKEN

1 First remove the feet, to make the bird easier to handle. Bend the legs and put a sharp knife through the joint.

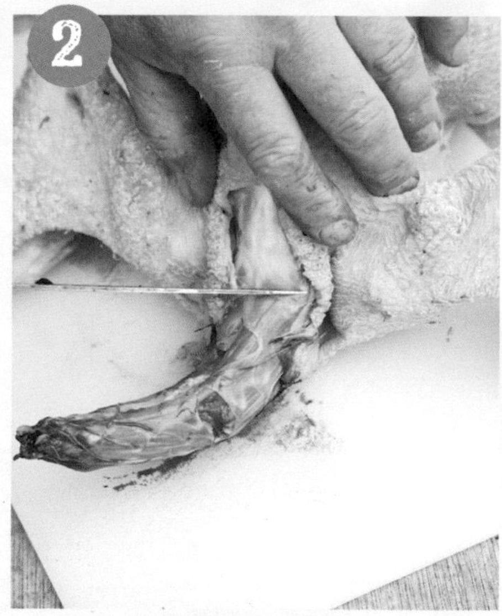

2 Remove the head where the vertebrae are separated, slice the skin along the length of the neck, then remove the neck.

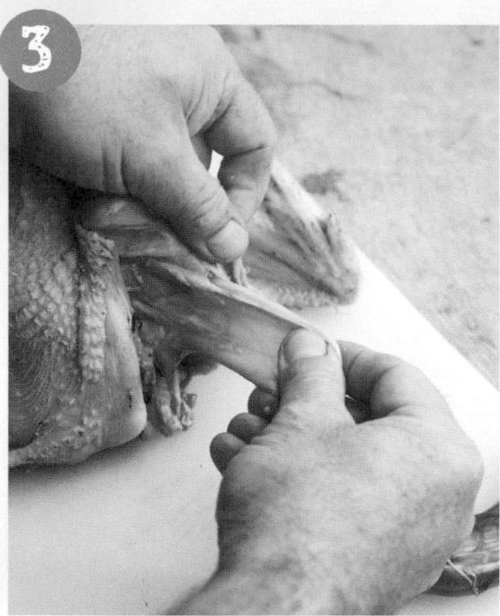

Loosen the crop sack from under the skin near the breast. This involves working your fingers between the skin and the crop.

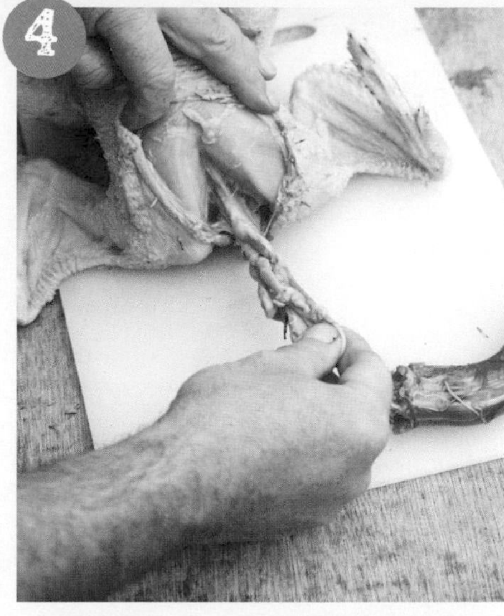

Separate the crop and tubes that go through to the body cavity, so that they can be pulled through when you remove the innards.

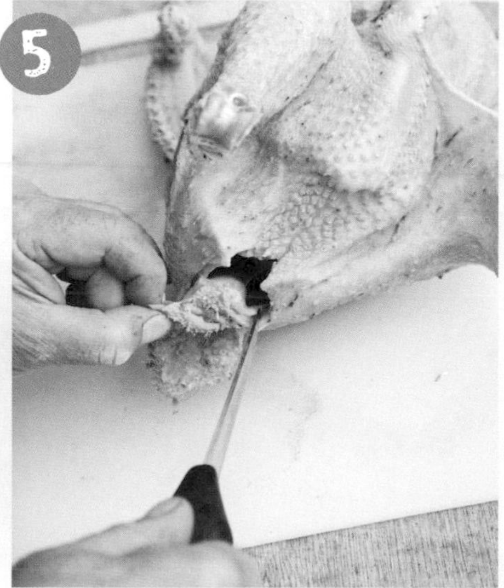

Cutting around the vent has to be done carefully so that you don't nick the connecting tube. Pull the vent out and increase the size of the hole so you can get your hand in.

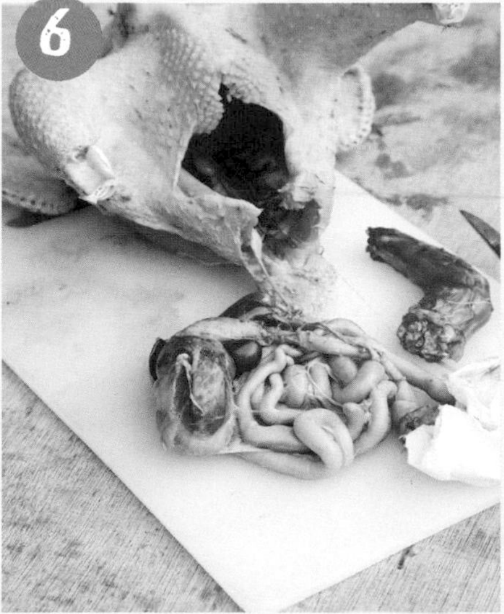

Wriggle your fingers up under the breastbone until you have all the innards under your hand. Draw them all out in one movement. Then wipe the carcass with a damp cloth.

2
DUCKS

INTRODUCTION TO

DUCKS

Why keep ducks? We decided to start keeping ducks when we moved to a place with running water on the plot. You can certainly keep them in a garden without a pond, but we felt that the stream provided the perfect opportunity to raise ducks in their natural habitat. Our motivation was primarily for enjoying their eggs – the large yolks have a rich flavour that makes them superb to cook with and excellent for baking – and the other reason we wanted to keep ducks was because we love eating duck meat and wanted to rear our own at home.

PROS

- Duck eggs are fantastic!
- They are very amusing birds, with real character.
- Keeping ducks avoids the guilt when eating commercially reared duck meat.
- Ducks will imprint on you if you rear them yourself, and this makes them very easy to put away at night.

CONS

- In spring the mating season makes the duck pond a place with a lot of noise, flapping of wings and chasing of females.
- Ducks are fairly messy, particularly in a small area, and duck poo is slimy – very easy to slip in if you're not careful!
- You really need to dig a pond if you want to have happy ducks.

BASICS

Ducks are hardy animals that will always find some food for themselves. We will often sit on a bench near our pond and watch as they waggle, quack and paddle around in search of tasty morsels. They do still require feeding, but on the whole they are a very low-maintenance animal to keep. The key to successfully rearing ducks is to give them regular fresh water and protect them from predators.

GUIDELINES

- Clean out the duck house once a week. It is vital to provide them with a clean bed – especially if that's where they are laying their eggs, too!
- Ensure that your ducks have clean water on a daily basis.
- Use your ducks as slug hunters in spring before you plant out any seedlings in the vegetable plot. Ducks are extremely useful animals to keep on your plot because they are selective eaters – they are less likely to scratch in garden beds than chickens, and can be useful for keeping down the population of pests like slugs.
- For the first few weeks, avoid letting ducklings go swimming, as they won't have developed their natural protection of oily feathers and could easily be eaten by predators.
- Keeping ducks in an area where you have lots of wildlife will significantly alter the natural balance. Ducks will eat all plants and

Space to roam means exercise – and that means good meat

Ducks need easy access in and out of the water

small creatures without leaving anything alive, and a stream could quickly have all its small amphibians, insects and wild plants eradicated.

- Male ducks will fight and gang up on females when mating. This is disturbing to watch and can be avoided by keeping fewer males in the same area as females over the spring mating season.
- Avoid leaving ducks with stagnant water and don't eat any eggs that have been laid in or close to dirty water.

ESSENTIAL EQUIPMENT

- Pond or large water container
- Corn in plastic bins
- Straw
- Sawdust
- Duck house
- Ramp for access
- Nest box
- Fencing

CHOOSING WHAT TO BUY

- Indian Runner ducks don't have much meat on them but are good layers. They have a distinctive upright stance and stroll around like penguins. On average you can hope for about 180 eggs a year.
- Aylesbury are large table ducks that are famous for being good to eat, and they can weigh up to 4.5kg (10lb).
- Welsh Harlequins grow into a good all-round bird. They're decent layers and are big enough to be an impressive table bird.
- Khaki Campbells make a great choice for egg production. They're made up from Mallard, Indian Runner and Rouen. Capable of about 300 eggs a year!
- Muscovy is a breed that we love. These ducks are heavy, placid and make excellent mothers. They are very good fliers, so you'll probably want to clip their wings. Be careful when picking them up, as they have sharp clawed feet. Muscovys are our favourite ducks for good eating and average laying.

FOOD

Ducks aren't grazing birds like geese but they will supplement their diet if you give them access to some land. They are also partly carnivorous and will happily eat slugs, snails, worms, frogs and other insects. Therefore in spring, before we commence the next round of crop planting, we allow our ducks access to the vegetable beds to find and eat any hiding pests. Don't get carried away and give them free range of your beds all year round, though, as they will cause damage to young brassicas and eat peas and lettuce. Keep your ducks in a fenced-off area with a pond, to avoid stepping directly in duck poo when you go outside in the morning. Feed your ducks corn, or feed specifically designed for ducks, daily to increase the number of eggs they lay and to fatten them up for the table.

WATER

Provide ducks with their natural environment where possible. This involves complete access to clean water. Flowing water or a pond are ideal, but if you don't have this you should ensure that you provide a deep container of regularly changed water. Ducks need to be able to periodically submerge their entire heads to clean their eyes and nostrils. Quite simply, if you don't have any natural water on your plot we would suggest you don't keep ducks. Ducks will drink water straight from a pond, so it is important to make sure that it doesn't ever turn stagnant. If you try to keep too many ducks in the same pond you'll find that they can make the surrounding area incredibly messy!

METHOD #8

DUCK HOUSES

A duck house is a basic structure and building one gives you the opportunity to play at being an architect for the day. We have made a host of designs over the years and each one is slightly different: some have sliding doors, others are on stilts in the pond, some even have their own house number next to the front door. Our advice is to get stuck in and build your own to save money and make something that is right for your location. Ducks aren't too fussy, and even if your carpentry skills leave something to be desired you'll find that your tenants won't complain! As long as you make sure that your duck house is easy to clean out and practical to collect eggs from, you can be as creative as you like.

PLANNING A DUCK HOUSE

Because of the oily coating on their feathers, ducks don't mind wet weather. However, it is important to provide a warm, dry space for them to sleep. Housing ducks is very simple. A conventional hen house can be easily adapted and placed near a pond.

There's no need for perches in a duck house, as the ducks will sleep on the ground, but you may want to provide a ramp leading into the house, as they tend to be fairly clumsy on their feet and can easily damage themselves. Additionally the housing should be rat-proof and fox-proof, draught-free but well ventilated and always have fresh, dry bedding.

If you are keeping a large number of ducks, using an old outbuilding or barn with a hole in the wall to the ducks' outside area is a good move. A cement or stone floor will be more convenient for cleaning, and a good supply of feeders hanging from the roof will provide even access to food. Install water containers, too, so that your ducks have access to water for 8–12 hours each day.

BUILDING A DUCK HOUSE

You will need some basic equipment to build a duck house. This includes wooden featherboard, plywood (marine plywood if available), roofing felt, wood screws, tacks and hinges.

To avoid rising damp and provide extra ventilation around the duck house it is vital to have it sitting off the ground. We use blocks of thick 5 x 5cm (2 x 2 inch) wood as a slatted base before building upwards. These bits of wood may need replacing after a few years, to extend the life of the duck house. A solid square of plywood set on top of these will serve as a good base that is easy to scrape clean. If building on water, sink the vertical uprights at least a third down into the silt base of the pond. Leave at least 30–60cm (1–2 feet) before attaching the base, to allow for changing water levels.

For the house itself, we tend to use a simple frame made out of 5 x 2.5cm (2 x 1 inch) timber and quickly clad it using short wood screws and overlapping featherboard. You can

weatherproof the wood to protect against rain, but don't paint the inside with any chemical paints.

The duck house must also be secure from predators and free from major draughts. Some ventilation is a very sensible idea, so drill holes in the walls under the roof of the house. Make sure that these are out of the prevailing wind direction and not big enough for vermin to climb through.

Make sure the door to the duck house is easy to open and close and that it is possible to close it securely. You want the size of the door to be 150% the size of your biggest ducks. We place a latch on some hinged doors or install a clever sliding system – like a miniature portcullis – so that gravity foils any foxy attempts to enter their home. To make this style of door, place 2 runners either side of the cut-out door hole and use a suitable-sized sheet of wood as the door that will slide up and down between them. Secure a hook at the top of the house for the sliding door to be held open during the day.

If you are going to finish off your door in style, install a small ramp with some slats for grip so that your ducks can make their way in and out of the house. This will avoid any clumsy injuries and prevent the ducks breaking a leg.

The top of your duck house can be constructed cheaply out of wood or corrugated tin. The down side to tin is that it is very cold in the winter and hot in the summer. This can lead to eggs going off more quickly when the weather is very warm. We prefer wood covered with roofing felt, but it's a bit more expensive. When attaching the roof, it is worth making it hinged so that you can easily lift it up for access to eggs or for cleaning. We also always put in a small swivelling support to hold the roof open (like under a car bonnet) – this makes the whole task of replacing dirty bedding with fresh sawdust much easier!

Jostling for position

A waterside residence

METHOD #9

DUCK PONDS

It is not vital to provide your ducks with their own pond, but it is lovely to watch them paddling around and diving under the surface to forage for things to eat. We believe that if you keep poultry, it is in your interest to provide them with a free-range area that mimics their natural environment. Ducks will remain cooler in warm weather, clean and in better all-round condition if they have access to water. The bigger the better with a pond, but make sure you prepare yourself for the inevitable – ducks will make your pond a mess!

PLANNING A POND

Ponds need water, and in that little statement lies the majority of the problems associated with making your own pond. Water is heavy and moving it takes a lot of energy, so when considering where to site your pond it makes sense for it to be positioned downhill from a water source so that gravity can do as much of the work as possible. You will find that the water you put into your pond will steadily be absorbed into the earth, unless you have clay soil, or you line your pond or you add more water than gets absorbed. We've solved this problem by having our stream feed through our pond.

CREATING A POND

Mark out the size and shape of your pond using a hosepipe or by drawing a line with a bucketful of sand. Then, using this as your guide, dig out the pond. A spade is all you need to dig your pond, but if you are making a very large one you might want to consider hiring a mini digger for the job.

Your duck pond doesn't have to be deep, and you should try not to think of it as a wildlife pond. There is unfortunately very little chance that frogs and water insects will survive the ducks' voracious appetite.

Therefore your pond can afford to be relatively shallow – 30–60cm (1–2 feet) is sufficient. Make sure the sides of the pond are shallow so that it is easy for your ducks to get on and out of the water.

Remove any sharp rocks from the base of your dug-out pond. Line it with a thick pond liner and fill with fresh water. Don't use a thin plastic liner, as it will quickly disintegrate and can become a hazard to your ducks and other wildlife. If you have an area of clay ground, you can create a natural pond without a liner.

The edges of your pond will very quickly start to be eroded by the ducks. They love to undercut banks when they are feeding, and the flow of water will also alter the shape of your pond. To keep the shape for longer, plant water-loving plants like bamboo and willow – they will thrive in the rich, damp area and help to hold the banks of the pond together. Other aquatic plants may also thrive, but they will only survive if they are not to your ducks' taste. Try a few sacrificial experiments before spending lots of money on expensive edible decorations.

Like a duck to water

Position large rocks and stones around the edges to preserve the pond liner and the integrity of the pond. However, always try to make sure that in at least a few places the edges of your pond are shallow. Ducks will struggle to get in and out if the sides are too steep.

LIFE AROUND THE POND

If you have the space, try to provide outside cover for your ducks so that they can hide from harsh weather: they need both shade for very hot days and shelter from heavy rain or snow. We also like to give ducks some fallen tree branches or old logs to perch on, and plant a mixture of shrubs and trees – using tree protectors when they are young. You will also need to fence around your duck pond if you want to protect ducks from predators.

ADDING FISH TO YOUR POND

If you want to really optimize your duck pond, you can use it for fish and ducks. There is a delicate balance that can be struck here, which works to your advantage. The duck poo will provide rich nutrients for the plant life, and fish such as carp or roach will feed on the smaller animals and organic material. The danger is too many ducks and not enough oxygen in the water. Worth considering if you like a symbiotic challenge.

Rats love to live by water

METHOD #10

PESTS & PROBLEMS

Ducks will be fairly resistant to smaller insect pests if you keep their house clean and give them plenty of fresh water. The key thing is to provide them with good living conditions and adequate fencing. We tend to feed our ducks by scattering food around their enclosed area, so you might think that rats would be a major problem. However, because we keep Muscovy ducks this rarely happens – they are the vacuum cleaners of the wildfowl world.

PREDATORS

Ducks are at risk from predators. Proper fencing is the best way to protect your birds. When ducks are small, protect them from above with cheap netting – large birds of prey will take your ducklings.

RATS

Rats are a constant issue with poultry keepers and dealing with them is necessary to ensure the wellbeing of your ducks. Rats are all around us, and in fact even in urban areas they are still never that far away. The normal balance of cat and dog control seems to work for us – they are the best deterrents. But sometimes we use other forms of pest control to significantly reduce the risk to our ducks.

For us, dealing with rats involves setting a series of live traps and sometimes putting poison in places where we know they hide. We see poison as the last resort, because it can sometimes get into the food chain and cause problems in other eco-systems. However, there are products on the market that reduce this risk, allowing the poison to be taken back to the rats' nest where it can kill more than one rat at once. It also happens to be the most effective method of control that we know. We generally only use poison when we have ducklings or chicks that are vulnerable. Keeping larger breeds like the Muscovy reduces the risk of rat attacks.

MITES

Ducks are susceptible to mites. Excessive scratching could mean a mite infestation so watch your birds each day for signs of illness and if necessary treat them with insecticide.

BOTULISM

This is a more serious problem and is caused by bacteria in rotting animal and vegetable waste. The toxins can leave a duck in great distress and with a serious loss of muscular

control. Avoid this by keeping ducks out of muddy, stagnant water – especially in hot weather. Ensure that there is a good supply of fresh drinking water for sick ducks. A traditional way of dealing with the problem is to add Epsom salts to their water (1 teaspoon to 600ml/1 pint).

LAMENESS

Ducks use their webbed feet to great effect when in the water, but on land they are not too sure-footed. Their legs are also quite fragile and susceptible to injury and long-term problems. A duck can develop lameness as a result of a bacterial infection from a cut or because of a strain from falling. Ways to avoid lameness are never to pick up a duck by its feet, to provide ramps for the duck house and to clean any cuts with disinfectant and then place the duck on clean straw for a day or two. Occasionally ducks can develop a limp because they are not getting enough vitamin B3 – you can remedy this by trying to feed them yeast extract on a slice of bread!

SLIPPED WING

The symptoms of this condition are that the large primary feathers on the wings of your ducks turn outwards. They look very odd – almost as though the wing has been bent the wrong way and broken. It occurs when ducks are fed a high-protein diet and grow too fast. The solution is to consider bringing in a different breeding stock or to avoid giving growing ducks too much protein in their feed while they're developing feathers.

FLYING AWAY

You may have to trim your ducks' wings so that they don't end up escaping to the local duck pond. At this point it's worth mentioning that domestic ducks that are 'released' to wild duck ponds rarely survive, so think twice before you emancipate any that you're tired of.

A natural mother

METHOD #11

LAYING & BREEDING

The rate at which ducks lay eggs depends heavily on their breed. The larger, slow-maturing birds are less likely to lay many eggs, but smaller ducks can lay more than some chickens. We are always astounded by the sheer quantity of eggs that our ducks give us -- especially considering they are more expensive to buy in the shops. But we're delighted to have them -- they're delicious.

ESSENTIAL EQUIPMENT

- Nest boxes
- Water container
- Feeder
- Rat-proof run

EGG LAYING

Generally a duck will start laying from 6–8 months old. Ducks lay best when there are a few of them living together. They like to have a sheltered and comfortable nesting area and will often return to the same spot day after day. Follow the same advice as for chickens (see page 28). Spring is the usual time for ducks to start laying, although sometimes you'll get eggs over the winter. Ducks, like chickens, lay more eggs when there are more hours of sunlight. The number of eggs corresponds incredibly closely to the longer days, and slows down when it gets dark earlier. You don't need a drake for the ducks to lay, but if you are planning on hatching ducklings he's fairly important! We like to keep a mixture of different breeds of duck –some for the table, some as part of our integrated pest control (basically for eating slugs) and others to provide us with delicious eggs.

BREEDING

Try to bring fresh blood into your breeding stock each year. This avoids any hereditary problems and gives you the chance to explore breeding potential. If you are breeding something like an Indian Runner duck, make sure that there are 4–5 ducks for each drake – they are randy little individuals. Sadly, ducks with the Mallard genetic background will forcibly mate with female ducks and during the mating season there is a risk that the females may die by drowning. This is difficult to prevent, other than by having a shallow pond and a good ratio of females to males.

BROODY DUCKS

When a duck goes broody she will puff up her nest with feathers and straw. On average ducks will happily sit on clutches of between 9 and 11 eggs. Broody ducks do not like to be disturbed, so it is a wise idea to provide easily accessible food and water. We also like

to protect them by making sure they are inside rather than nesting outside, where a predator could get them at night. Incubation normally takes 28 days. Each year we are amazed when one morning we go to let out the animals and a procession of fluffy ducklings follows their proud mum.

CHICKEN BROODING

Fertilized duck eggs can be placed under a broody hen and she will incubate and hatch the eggs as if they were her own. Hens are much more robust than ducks as egg sitters when they go broody. They will get grumpy if disturbed, and you can expect to get pecked if you check under the bird and try to move it aside; however, they will settle back again without considering giving up and deserting the eggs. It doesn't take long to discover which hens make the best mums, they will have successfully reared a clutch and they'll let you know how grumpy they are!

METHOD #12

INCUBATING & HATCHING

The process of incubating and hatching duck eggs is similar to that of chickens, but the eggs are bigger, so they need adjusting on the rails in your incubator and they take slightly longer to hatch. We tend to allow the ducks to rear their own ducklings, because newly hatched ducklings will imprint on to people and it all gets a bit confusing when it's time to learn to swim.

ESSENTIAL EQUIPMENT

- Thermometer
- Spray bottle
- Incubator
- Torch
- Duck crumb
- Incubating bulb – infra-red
- Brooder

FERTILIZED EGGS

Choosing healthy eggs that have a good chance of success is the most important part of the incubating process. Select eggs carefully by inspecting and candling them (see pages 30–31) at the time they are put into setting trays. It's not worth setting eggs that are even slightly cracked, double-yolked, misshapen, oversized, undersized or dirty.

INCUBATOR

Duck embryos need warmth to encourage development at a natural rate, and achieving the optimum temperature is vital. This is why ducks will leave their nest, cool their feathers in water and return to the eggs. Replicating the same environment is the success story of modern incubators.

Humidity levels need changing at certain stages of incubation. At the beginning they stop the shell becoming too dry and help with the necessary weight loss from the egg. Later they are very high and keep the shell soft, enabling the duckling to force its way out into the world.

Eggs from ducks like the commercial breed of Pekin require 28 days to hatch. Eggs from Muscovy ducks hatch slightly later, at 35 days

after setting. Start the incubator and allow the temperature and humidity to stabilize for a day before setting the eggs in it. Set the temperature at 37.5°1C (99.5°F) and the humidity at 55%. Where available, set the ventilation as recommended by the incubator manufacturer.

Eggs must be turned, either automatically or by hand, a minimum of 4 times a day. Fortunately, modern incubators normally have a mechanism that turns the eggs hourly. This ensures that the eggs are evenly warmed and that the membrane inside doesn't stick to the shell wall.

For best results, set eggs within 3 days from the time they were laid. Always set eggs with the small end pointing down. You will find that checking your incubator frequently to make sure it's working properly is worth the effort – especially on the first day. Continue checking thereafter 4 times a day.

CANDLING

Seven days after setting you will need to candle the eggs and remove any that are infertile – either clear or cloudy.

HATCHING

After 25 days (dependent on the breed of duck), transfer the eggs to hatching trays. These are sometimes within the same incubator, or in a separate machine called a hatcher. Candle once more and remove any eggs with dead embryos. Set the temperature of the hatcher to 37°C (99°F) and the humidity to 65%. When you see or hear that hatching has started, increase the humidity to 80% and increase ventilation openings by about 50%. The first sign is usually a faint pip noise from within the egg, followed by a tiny knocking on the wall of the shell. As the ducklings make their way out of the shells, gradually lower the temperature and humidity so that by the end of the hatching the temperature is at 36°C (96.8°F) and the humidity is at 70%. Vents should be opened to their maximum setting by the end of the hatch. Remove ducklings from the hatcher when they are dry.

BROODER

After ducklings have dried off the incubator, transfer them to a brooder as you would for chickens (see page 31). Ducklings will grow faster than chickens and need to have more water available to clean their eyes and feathers. They are also more messy than chickens and will need their bedding replaced more often. You can take them for walks outside when they are 5–6 weeks old and they will imprint on you (see page 73).

METHOD #13

SLAUGHTER

Gently does it

Larger poultry have thick necks that are not as easy to break as a chicken's, so their slaughter requires a slightly different approach. It is possible to kill a duck on your own by putting a broom handle over its neck, standing on each end and jerking up the body; however, we prefer a team effort as this is swifter, more controlled and therefore more humane. If your duck is small, it can be slaughtered like a chicken (see pages 32–33); however, medium to large ducks need to be killed using a simple device that holds the duck in place so that its neck can be broken.

PREPARATION

As with chicken, you should prepare to kill your duck in the morning, when the digestive system is empty. As a table bird you can expect a large duck to weigh upwards of 2.5kg (5lb). Handling needs to be calm, but the birds are powerful and have to be firmly held.

We have a simple contraption that holds the duck in place; it's an upturned traffic cone, mounted in an old stool. The duck's neck fits through the hole at the bottom but the shoulders cannot be pulled through, which means the neck can be stretched. Any form of cone can be utilized – you just need to ensure that there is sufficient room below to break the neck and bleed the bird. To break the neck we have 2 broom handles tied together at 1 end.

With 3 people, we can keep the operation as smooth as possible; 1 person holds the duck's legs and the other 2 break the neck using the broom handles. Only when you are ready should you collect the duck for slaughter. It is essential that everyone understands exactly what they need to do before the duck is picked up, as any delays after this would distress the bird.

HOW TO SLAUGHTER A LARGE DUCK

Turn the duck upside down and feed the head through the hole in the bottom of the cone – you may have to reach up and pull it through.

Put the broomsticks on either side of the neck, ensuring that there is enough ground clearance for you to stretch the neck fully.

When ready, squeeze the handles together tightly at the end that is not tied together and bend the head back, while pulling down until the neck breaks.

When the neck is broken, cut the throat and allow the bird to bleed out. Then wrap the throat with kitchen roll and use rubber bands to hold it in place (see page 33).

In the end, it's just about cooking

METHOD #14

PLUCKING

It may seem like a silly thing to say, but there is an awful lot more plucking involved with a duck than a chicken. It's not just because they're usually bigger, but because ducks have a second layer of insulation – the down – to provide them with buoyancy and heat retention. When you set about plucking a duck, be prepared for a vast quantity of feathers to be produced – the down gets everywhere, so expect to be covered in a fine layer by the end of the plucking process. Duck down is renowned for the warmth it brings to duvets and pillows, so consider collecting the feathers for use. Once you have plucked your duck, draw it in the same way as you would a chicken.

HOW TO PLUCK A DUCK

Tie a rope around the bird's feet and hang it at a convenient height. The wing feathers of the duck are hard to remove by hand, so use pliers and pull the feathers out individually.

Make short, sharp downward tugs, taking a small clump each time, to remove the outer feathers in the first instance.

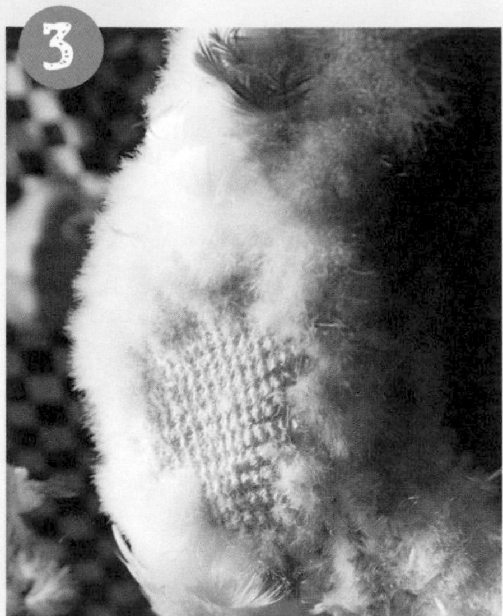

Next, pluck the down feathers in the same way as the outer feathers, but these can be saved. Expect to see some small, wispy hair-like feathers left over.

Loosen one leg from the rope to pluck around the tail and the vent. Singe the fine feathers with a blowtorch or a naked flame. Move swiftly so that the flesh does not heat up.

Rub the singed feathers off the carcass with your hand.

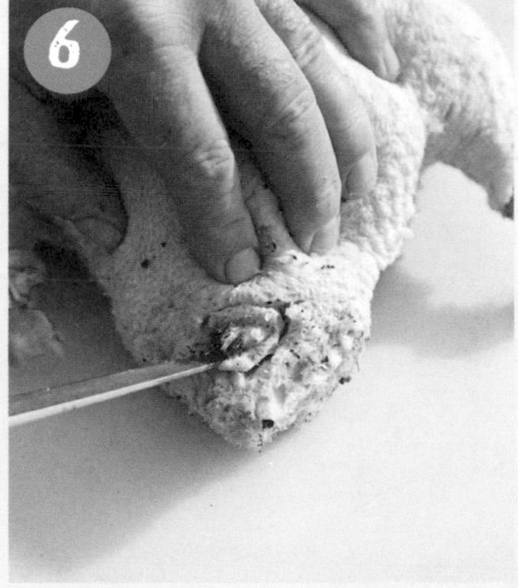

Using the point of a sharp knife, remove the musk gland at the tail (on the underside of the parson's nose), then draw in the same way as a chicken (steps 1–6, pages 36–37).

GEESE

INTRODUCTION TO

GEESE

Why keep geese? Keeping poultry tends to not be the most glamorous lifestyle choice. However, when it comes to geese there are not many domestic animals that are as splendid. Geese are amazing to rear yourself, and demand a degree of respect due to their size and impressive wingspan. We find them hardy, tough and surprisingly self-reliant. Furthermore, we like the fact that our geese are multipurpose birds. They are first-class grazers and excellent guard animals, and they usually start laying enormous eggs in early spring. One thing to bear in mind is that geese can live to 25 years old and are a real commitment – unless, of course, you are rearing them for the table.

PROS

- Geese lay huge eggs – when boiled for breakfast you need a whole regiment of soldiers to dip in them!
- Geese are great animals for keeping the grass down in an orchard. Geese eat grass like lawnmowers and can even be used for weeding between strawberry plants or vines, as they don't like eating broad-leaved plants.
- You can pay a huge amount of money for a goose to eat, but rearing them yourself works out as affordable.
- Geese are quite resistant to predator attacks.

CONS

- The honking noise of a goose can be rather off-putting. It's what makes them great guards, but sometimes it can shatter the illusion of a peaceful day in the quiet countryside.
- They're big birds to handle if you need to clip their wings. Respect them when you pick them up or you will receive a strong wallop from their wings.
- Grass goes straight through geese, which means that wherever you have geese you will have lots of goose poo.
- They have a decent nip on them if they are in the mood for an argument – they require you to stand your ground and let them know who's boss.

BASICS

Successful goose keeping starts with choosing the right breed for you.There are a staggering variety of different geese, all with their own special appeal. You can buy them as egg producers, table birds, lawnmowers and/or guards. It's important that you keep a breed that you like, one that makes you smile, as they can live for 25 years.

GUIDELINES

- Make sure you imprint yourself on goslings when they are young by being around when they hatch and over the first few days. This will make it much easier to put them away at night when they grow up!
- Supplement their diet with corn.

A gander and 3 geese wander around as a 'gaggle'

A gander is a substantial bird and best not riled

- Make sure their water is replaced daily.
- Clip their wings every few months or you could find that your gaggle sees some tourists overhead and flies off to join them.
- Try not to break up a mating couple. Geese pair for life, so we always keep one breeding pair to supply us with next year's goslings.
- Never leave any netting, loose wire or string in their enclosure. They can eat it or strangle themselves.
- In our opinion it's not worth risking leaving geese out overnight. Despite being big, at night they are still vulnerable to fox attacks.

CHOOSING WHAT TO BUY

- Pilgrim is a lovely breed of goose with an unusual sex-linked plumage that makes it easy to distinguish between genders. The gander is white while the goose is light grey.
- Brecon Buff is a lovely sandy-coloured goose. Not only does it look lovely, it also tastes great!
- Embden is an excellent and very large table breed. The Common English is often an Embden crossed with a Toulouse.
- Roman is particularly good if you want a goose that matures quickly and can be killed at a young age – still with plenty of breast meat.
- Chinese geese are excellent egg layers and look very graceful. They are also good foragers and can be ready to slaughter as early as 8 weeks old. The meat is a bit darker than that of other geese but not as greasy.

FOOD

Keeping geese requires more space for grazing than ducks or chickens. A small orchard or paddock is perfect for them. Try to avoid long grass, as this can cause a problem called crop-binding. Scythe or mow the grass so that it's approximately 10cm (4 inches) tall. The geese will then take over mowing responsibilities. The more free-range they are, the less you will need to spend on feed. We give our geese corn each day but they seem to much prefer the grass. Geese that are fed on good short grass should be fine if they are given less corn over the summer months.

WATER

Although they are officially classed as a type of waterfowl, geese don't absolutely need a pond – all they require is access to clean water so that they can keep their nostrils and eyes clean by immersing their heads. We have found that a large container will do until you can make something a bit more permanent, like a small pond, but only as long as it is at least 30–60cm (1–2 feet) deep and you change the water regularly. The difference between a pond and not a pond is that the geese will stay much cleaner and reduce the chances of parasitic insects or disease.

GREEN GEESE

In the past, rearing geese was linked with the annual growth and decline of grass. This old method makes good financial sense and reduces the amount of time needed to mow the lawns. Goslings begin grazing when the grass is fresh in mid spring, and when the grass slows down around mid autumn we slaughter them. It is a harsh natural cycle, but incredibly cost-efficient. However, you can still supplement their grass diet with the occasional corn feed if you want to fatten them up and keep them healthy.

REARING GEESE FOR MEAT

The methods for slaughtering, plucking and drawing geese are the same as for ducks, so see pages 54–57 for advice on these topics.

METHOD #15

GOOSE SHELTERS

As well as being intently curious, geese also seem to be fearless. Our geese will often chase away our dogs if they stray into their enclosure, and they love to attack the wheelbarrow when we are cleaning out their bedding. However, we have always provided our geese with a covered shelter to provide them with privacy when laying eggs and to protect them from bold predators.

PLANNING A SHELTER

Geese are very hardy birds that don't get ruffled even when it's snowing. A goose shelter is important so that they have a good place to lay eggs and as a protection from predators rather than from harsh weather conditions. We have always chosen to reuse a large, old shed for a goose house because they are readily available in local papers or advertised in poultry magazines. The key is that there should be at least 1 square metre (10 square feet) for each bird. The height should be above the birds' head height, and the larger the pen the easier it will be to clean out and collect eggs.

Waterfowl can be kept in standard poultry housing, but heavy domesticated breeds prefer the entrance and exit to be at ground level – they find it harder than chickens or ducks to jump up to an entrance or walk on ramps. Installing a solid floor for the goose shelter not only makes it easier to sweep clean but also enables you to keep it dry. A dry layer of bedding will prevent the birds from getting arthritis in their feet. Use layers of cardboard, newspaper or straw and replace when damp. Geese, like all waterfowl, make a fair amount of slippery poo, which is messy but great for the compost heap.

POSITIONING A SHELTER

A goose shelter should be situated in a large area that is close to a water supply. It needs to be surrounded with netting or fencing to protect the geese. Geese can be noisy, so site their shelter away from your house, and those of your neighbours, to reduce the noise pollution.

BUILDING A SHELTER

Erecting a shelter for your geese can be cheap and easy. Straw bales are a perfect base for your structure and can be composted when, or if, your geese are slaughtered. Construct the walls using overlapping bales, as you would if you were laying bricks, providing enough floor space per bird.

A STRAW BALE SHELTER

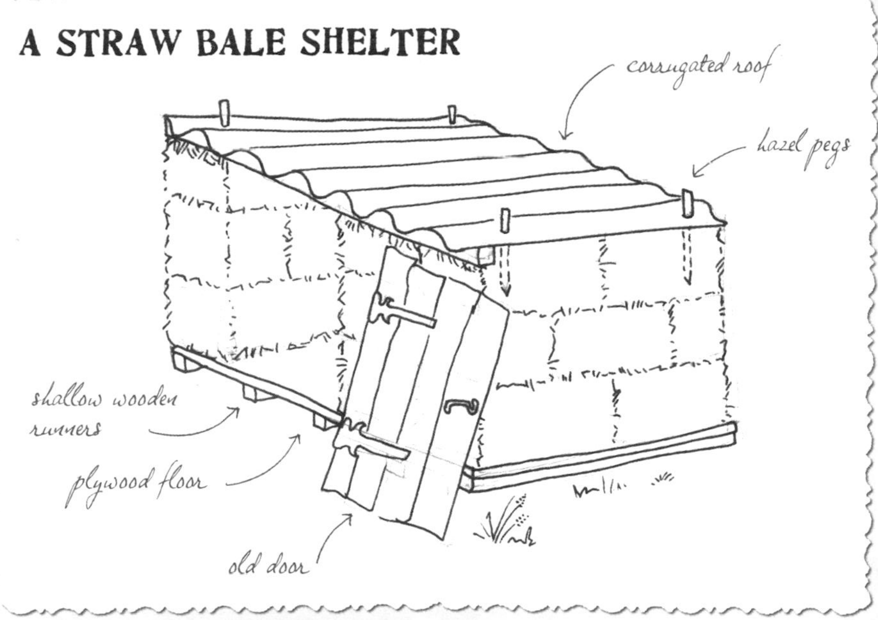

Then on top, at an angle sloping towards the back, simply place a wooden or corrugated metal sheet roof. You can secure this roof by using hazel spikes, which you cut to approximately 60cm (2 feet) in length, sharpen and drive down into the bales. This will create a hut in which the geese can live and lay their eggs.

Each night when you put your geese away, position an old wooden door or a wire grill in front of the gap to stop any predators getting in. If opting for this low-impact shelter it is still worth placing a solid floor inside for health reasons. A standard garden shed with added ventilation holes drilled into the walls makes a good alternative.

GOOSE HERDING

Putting geese away every evening may at first seem like a nightmare chore, as geese can be difficult to herd. Yet, after some patient practice you will find that they can be easily trained. Use a walking stick as an extended arm to guide your flock of geese towards their house. Gently guide them in, using a familiar sound that they will quickly learn means 'bedtime'. Geese are very sensitive to unusual changes in their routine, so try to keep a regular time and approach for putting them away – varying it according to the changing seasons.

CLEANING OUT THE SHELTER

When cleaning out your geese it is worth using the straw and poo mixture as a mulch around the base of your fruit trees. The nutrients from the poo will feed into the surrounding roots, boosting the trees' growth, and the straw acts as an excellent mulch to reduce weeds and keep the base of the tree warmer in winter. Just remember to leave a small gap so that there is no direct contact with the trunk.

Edible lawnmowers

METHOD #16

GRAZING ORCHARDS

Geese are grazing birds but they love to eat more than just grass. By planting an orchard in their enclosure you will create a really good working relationship between the trees and the geese. The geese will keep down the grass and reduce the growth of other competitive plants around the trees. Additionally, by keeping the grass shorter the geese reduce the population of any burrowing animals that enjoy munching on the roots of fruit trees. There is something classic about geese waddling in a line through an orchard -- a taste of the past, the perfect integration of productive plants with domestic wildfowl.

PLANNING AN ORCHARD

In an orchard you can grow anything you want, but we'd suggest choosing some productive fruit trees. Look for varieties of tree that suit your local environment. Talk to staff at your local garden centre, or simply look around and see what is growing well in your area.

A small orchard will provide you with all sorts of different fruit as well as a lovely shaded area for the geese. Two good choices for an orchard are free-standing apple and pear trees – they are really easy to grow and require little work. They should be planted between late autumn and late winter. Apples and pears like annual heavy mulching with rich compost. Pears like to grow in a more sheltered spot than apples and are a bit more fragile. They both benefit from a generous mulching of well-rotted manure, but remember to leave a space so that the base of the stem is not covered.

Planting fruit trees is an investment for the rest of your life. Once established, you will have a regular supply of fruit every year.

POSITIONING AN ORCHARD

Choosing the right location to site your new trees is extremely important, and the old test of knowing which direction is south is vital. Most trees will do best in a sunny spot, and some fruit will just about deal with shady conditions, but on the whole you should avoid planting under already established trees. Wind damage to fruit trees can be a nuisance, and planting your orchard in an overly exposed position can deter pollinating insects, so try to plant your trees in a sheltered area. Finally, try to plant in well-drained soil and avoid frost pockets.

PLANTING AN ORCHARD

After you have chosen a good site for your orchard, it's time to move on to the fun part! Planting a tree can be a serious shock to its roots, so it's best to wait until the winter months – when the plant is dormant and before the sap starts rising inside the trunk. Your geese may well be very alarmed by the whole activity, but it won't be long before they look at home.

To plant a tree, first dig a much bigger hole than the size of the root ball. Then drive a supporting stake down into the bottom of the hole. Put the tree in the hole, making sure that it is planted at the same depth as it was in the pot. Mix the soil you dug out of the hole with some rich compost and put it back in the hole around the root ball, firming it down around the tree as it goes in – do this gently, so as not to damage the tree roots. Build up the soil above ground level and then compact it down with the heel of your boot – this is called 'heeling in'. When you have finished, water the tree extremely well and put some more good organic mulch (ideally from the goose house) around the base of the tree.

PROTECTING SAPLINGS

Geese are curious and will try eating just about anything at least once. Young trees are particularly susceptible to having their bark nibbled, which can let in disease. To prevent this, put a tree protector on the base of the tree trunk for at least the first 6–12 months after planting. After that, keep an eye on the trees and protect as necessary.

MULCHING

Mulching is a key bit of maintenance when growing trees. To get your orchard to thrive and provide you with lots of fruit, keep the area around the base of your trees free from weeds and consider laying down cardboard or newspaper under a layer of hay or straw. Luckily, dirty bedding from your goose house is perfect to use as a thick layer of mulch!

CREATING A POND

Geese are fine with just enough water to drink and to clean their beaks and eyes. Nevertheless, giving them a pond is a very good idea if you have the time to make one. Geese enjoy cleaning themselves, and supposedly it helps with buoyancy for successful mating. Even better is if you have a small stream on your property or access to running water.

A good pond will mean more successful breeding

METHOD #17

PESTS & PROBLEMS

Geese are hardy birds that are usually problem-free. Prevention is better than cure, and the key to looking after your geese is to provide them with a healthy environment so that few issues arise. Make sure their area has no wire, string or sharp objects lying about and that there's good access in and out of a pond. We go out for a little walk a couple of times a day to look at the geese. This is an opportunity to see what they are doing and take notice of their daily routine. Often if there is something wrong their normal behaviour will be affected. The better you know your geese, the easier it will be to spot any problems.

WORMS

Worming your geese may be necessary if they have been stressed out by a long journey or have suffered from some other problem. The signs of a serious worm infestation are that the bird will lose weight and may move slowly. Ducks and geese can get gizzard worms, an unpleasant parasite that burrows into the gizzard, causing considerable damage and distress. Geese are more likely to suffer from its effect and die, so it is always worth worming geese when you buy or sell them.

PREDATORS

Hawks, mink, rats, stoats and foxes are all risks to newly hatched goslings. However, after no time at all they will have grown big enough to be far less vulnerable. The key thing is to have proper netting around their house to protect them when they've just hatched and trust your adult geese to look after the rest. Keep feed in secure bins rather than loose on the floor to avoid attracting vermin such as rats.

LAMENESS

Lameness is more common in birds that do not have access to water and are kept on hard surfaces. Calluses can form on their feet and an infection from a cut can quickly turn into a longer-term problem unless it is treated. A can of spray-on iodine is ideal for dealing with poultry and means you can quickly and

calmly deal with any minor injuries. Check for thorns or cuts, and if symptoms persist, separate the bird from the others and provide plenty of fresh bedding and food. Damp conditions in the area where the geese sleep can also eventually lead to foot problems. The answer to this is to clean out the sleeping area regularly.

ASPERGILLOSIS

Damp bedding and the use of hay rather than straw can lead to the growth of fungal moulds. These can damage breathing, lead to weakness and sometimes cause sticky eyes. Unfortunately, when a goose inhales fungus spores it damages its lungs and the infection is not treatable. The key here is not to have damp bedding and to make sure that their food is stored in a dry place so there is no mould growing in it.

FLYING AWAY

Wing clipping is important if your geese look as if they are going to fly away, but for the majority of domestic breeds it's not really necessary – they are too heavy to take off!

A goose is a difficult bird to catch. The way we deal with the challenge is to pick them up in the morning or evening, when they are in their house. The best approach is to grasp them firmly around the neck, behind the head – you want to avoid being bitten. A goose bite will rarely draw blood, but the twist and nip can be painful. You're unlikely to be attacked unless it's the mating season or there are goslings around. The solution is to stand your ground and react quickly. Pick up the bird by sliding your free arm under the breast and firmly grabbing the feet. Cradle the goose so that its head is behind your back and its bottom is facing forwards. This way there's no risk from either of the dangerous ends ... Use your arms to stop any wing-beating, as if panicked the bird can actually break a wing. This is another good reason to remain calm and not stress out your goose.

Never clip wings when the geese are moulting. Cut a line with a sharp pair of scissors carefully across one wing, removing only the primary feathers and avoiding cutting back too far.

METHOD #18

LAYING & BREEDING

Geese form a very strong family unit. A gander can live to 30 years old, and pairs mate for life. The result is the perfect animal to allow to hatch naturally. The down side is that every year at the same time your geese will become aggressive, loud and territorial. Collecting eggs is a mission that requires speed, agility and some real concentration. Bombarded with hissing, you will reap the rewards: delicious eggs with great flavour or very sweet goslings that will make everyone smile.

ESSENTIAL EQUIPMENT

- Nest boxes
- Water container
- Feeder
- Rat-proof run
- Bravery pills

EGG LAYING

A goose doesn't need a gander to be able to lay eggs. What she will need is a comfortable nest box with plenty of fresh straw. Unlike chickens, a goose will make her own little mound of straw to lie on, and when it comes time to incubate a group of eggs she will tuck them into the straw to keep them warm. The downside, if you want to eat the eggs, is that this means poking around in her nest to find them.

It's important to collect eggs properly so that stray ones don't attract vermin. It is also worth warning you that collecting eggs from a grumpy goose or an overprotective gander is a bit like running the gauntlet. We like to take along the trusty goose-herding stick, just in case. Take care – particularly when your head is between your knees looking for eggs and your rear is exposed to a goose attack.

Unlike ducks and chickens, geese won't lay throughout the season. They start in early spring and cease in mid summer. Normally you'll get a couple of clutches of eggs before the goose stops laying. The biggest difference, though, between geese and other poultry is that goose eggs are massive! Weighing approximately 200g (7oz) compared to a hen's 60g (2¼oz) – it's no surprise they lay fewer eggs. If you want to encourage your goose to keep laying and hatch goslings, leave them under her. Provide close access to food and water and wait for 30–32 days.

BREEDING

Hatching goslings naturally is far easier than incubating. Geese in their first year will instinctively try to hatch goslings if there is a gander with them, and it's a lovely thing to see succeed. Try to keep the size of the clutch of eggs down to about 6. This will mean that there is plenty of room under the goose and all the eggs will remain warm.

BROODY GEESE

Your goose will pluck down feathers from her chest to bulk out the nest and will constantly make little adjustments to her design. It is worth placing food and water closer to the nest during this period as she will not want to leave it for long. The average incubation period is 30–32 days, and once hatched it's normal for the goslings to remain under their mother for a day or two, surviving on the remaining food in the egg.

The way geese look after the goslings is lovely. They do not nurture in the same way as chickens; teaching their offspring the ropes is all done by example. Even as day-olds the goslings will peck at small green shoots of grass. We always feed them a dry supply of chick crumb and sometimes tear up bits of lettuce or tender green vegetables as a treat. Parent geese are very protective, so treat them warily when you herd them in and out of their houses.

Young birds will find obstacles where grown geese have no issues, so check your pen to ensure it is gosling friendly. Access to houses and water should be a slope rather than a step and don't forget that a gosling can slip out of a small hole and struggle to find a way back.

Not the planned nesting site!

METHOD #19

INCUBATING & HATCHING

Geese start producing eggs early in the year, around early spring and through until mid summer, laying around 30–50 eggs per season. That may not sound like a lot of eggs, but they are huge! Hatching your own goslings is one of the most fun jobs associated with poultry and wildfowl. This is partly because they will go through amusing stages – from cute and fluffy, to grey and awkward, to mature and impressive.

ESSENTIAL EQUIPMENT

- Thermometer
- Spray bottle
- Incubator
- Wire-wool sponge
- Torch
- Goose and duck crumb
- Incubating bulb -infra-red
- Brooder

FERTILIZED EGGS

Goose eggs must be collected as soon after laying as possible, and it's a good idea to write the date of collection on the egg shell with pencil. Before going into the incubator they need to be stored in a cool room (between 7°C and 10°C/44°F and 50°F), broad end up, and ideally turned once daily.

Dirty eggs need to be cleaned quickly with wire wool. Apparently any dirt on the eggs can seep through the shells and infect the embryo inside, which is not good. Gently scrub the dirty spots with a wire-wool sponge. This technique is better than washing, because the egg shells are porous, so washing can sometimes push dirt into the pores of the egg. For the best hatching results, eggs should be no older than 7 days.

INCUBATOR

Eggs should be positioned broad end up, or pointy end down. The incubator should be positioned in a place where the sun's rays will not shine directly on to it. A good temperature for the incubator room is a constant 18–20°C (64.4–68°F).

The temperature of the incubator needs to be between 37.5°C and 39.4°C (99.5°F and 103°F). The temperature inside the egg should be 37.7°C (99.8°F), which is why the temperature surrounding the eggs needs to be slightly higher. We set the temperature of our incubator at 38°C (100.4°F). It is also a good idea to have a thermometer outside the incubator. If the room temperature drops overnight, the incubator should be placed in a cardboard box, which helps to maintain a constant air temperature around it. Even a drop of a few degrees overnight can mean the difference between good hatching results and bad.

It is also very important to maintain a steady level of humidity. Goose egg humidity is between 45% and 55%. It's fairly difficult to maintain the humidity at a constant temperature, but it helps if you squirt the inside of the incubator with warm water.

It can be helpful to keep a record of the temperature and humidity each morning and evening, in order to have a record in case you have a bad hatching result. However, we find that the goose eggs read the same temperature and humidity each day, so we trust in the process.

If you have a modern incubator, it is unlikely you will need to worry about turning the eggs, because the incubator will do it automatically. They need to be turned at least 3 or 5 times a day. It has to be an odd number so that the eggs don't rest for a long period overnight on the same side each time (which they would do it they were turned an even number of times). It's advisable to mark the eggs with an O and an X on opposite sides, so you can easily keep track of which side to turn them on to.

After 14 days of incubation, the eggs must be sprayed with warm water at 39°C (102°F) immediately after turning. Goose eggs need to be incubated for 33–35 days before they hatch, and it is quite common that not all goslings will hatch at the same time.

Fair enough – they are ugly

CANDLING

Seven days after setting you will need to candle the eggs and remove any that are infertile – either clear or cloudy.

HATCHING

Follow the same approach to hatching as you would with a duck egg (see page 53); you will know when they are on their way because you will hear a quiet sound like a whistling kettle coming to the boil.

BROODER

Once hatched, leave the goslings to dry in the hatcher for 24 hours before moving them to a brooder with a heat lamp, chick crumb, water and sawdust bedding. Make sure they are secure and that the brooder is rat-proof.

IMPRINTING

The thing that geese see most in the 24-hour period after their birth is what they will think is their mother. Imprinting is fun in that the goslings are attentive and will follow you around, galloping after you. The difficulty comes when you leave them and that can cause them stress along the lines of 'where's mummy?,' so on balance we feel it best to leave them together as a group.

TURKEYS

INTRODUCTION TO

TURKEYS

Why keep turkeys? When we think of turkeys, its not eggs that spring to mind, it's a celebration meal with a roasted turkey as a centrepiece. These solitary woodland birds are not well suited to mass production, and those raised to the highest welfare standards tend to be expensive. It was to overcome the expense of organic turkeys at Christmas that we decided to keep our own. We know exactly what goes into each bird and the quality of the meat is amazing; they taste great. There is also the added advantage that they make very special presents for friends and neighbours.

PROS

- Turkeys are very gentle birds – probably a factor that led to their domestication.
- If you want a bird for a celebratory meal, there is something very special about knowing where it has come from.
- Turkeys are relatively quiet birds – the stags make a unique 'gobble' noise, but on the whole they aren't noisy.

CONS

- Turkeys are prone to illness and have a fragile disposition. Despite all your best efforts you may lose one from time to time.
- They are not very good at seeing in poor light, and putting them away in the evenings can take some patience. Our advice would be to put them in their house before it gets dark.
- You can't keep turkeys on the same patch of land 2 years in a row, so you will need to have enough space to rotate them.

BASICS

Keep your turkeys outside, in a small paddock surrounded by electric fencing to keep out any hungry predators. Give them a decent-sized shed to live in and install a big perch in the middle of their enclosure, on which they will tend to sit. Keeping turkeys requires very little work other than feeding and cleaning them out regularly. We feed our birds with turkey growers pellets that have less protein in them than commercial turkey feed. We are able to get away with this type of feed because we usually get birds as early as possible in the season; they have plenty of time to get big naturally and obtain a great deal of their protein by snacking on the insects and bugs in their enclosure. It is also vital to provide a good source of grit for your turkeys to eat. This acts as a sort of knife and fork for the birds, enabling them to mash up their food in their crop and digest the goodness.

GUIDELINES

- Always provide clean water for your turkeys to drink.
- Provide grit.
- Protect your turkeys with high electric fencing – this will keep foxes out and your birds in.
- Provide heavy-duty perches for them in their houses.

Amorous stags' wattles and snoods become very bright and impressive

Turkeys are inquisitive and don't tend to be timid

- Do not keep turkeys in the same area as chickens.
- Rotate the living space of turkeys each year and always avoid keeping them on the same land.

CHOOSING WHAT TO BUY

- Norfolk Black turkeys are our favourites and have striking black plumage. They are slower to mature than many commercial breeds but have a really tasty flavour. Although slow to mature, the stags can still reach a good 13kg (28.5lb)!
- Bronze turkeys are named after the unusual colour of their feathers, which are a green-copper brown in the sunlight. Bronze turkeys often require artificial insemination to reproduce. We have found them a calm bird to keep and one that can grow very large indeed.
- The Narragansett is an extremely tasty bird that is more hardy than most other breeds due to its mixed genetic selection. Narragansett turkeys are traditionally known for having a calm disposition and for making good mothers.
- Buff turkeys have reddish-buff feathers and the hens can be good egg producers. The added advantage of a Buff turkey is that the pinfeathers are nearly white, so it looks great when plucked.
- Bourbon Reds originated in Kentucky and Pennsylvania and can grow to a monstrous 15kg (33lb). They are a reddish chestnut colour, with white tail feathers.

FOOD

Turkeys need to be fed specific turkey food to ensure they remain healthy and get the necessary minerals they require. Free-range turkeys with access to a field will have a more balanced diet than those simply kept in a barn. The benefit for you is that they will also gain more protein from the land and hopefully grow larger. We use a galvanized feeder and top it up each day with fresh, dry turkey pellets. For the first 5 weeks they are fed purely on starter crumb and from 5 to 8 weeks this is phased out, by mixing in some growers pellets to bulk up their diet. Growers or finisher pellets have less protein but will form the turkeys' standard diet until slaughter. The reason we provide fresh food each day is that turkeys are messy creatures and will stand on the feeder or poo on their food – clean this out often, and try to avoid spilling food on the floor and attracting rats. We keep the feeders under cover so that the food doesn't get ruined by rain.

WATER

Turkeys need clean water so that they don't get dehydrated. Water also contains valuable trace minerals that will help with their growth development. Again, topping up their bowl with fresh water is a daily job.

CLIPPING WINGS

If you want to restrict the movement of your birds, clip one of their wings. Clip both wings and very quickly they will be flying away again; one wing makes it much harder for them to lift off. Simply cut their primary feathers off with a sharp pair of scissors where they line up with the next layer. Be careful not to cut too close to avoid them bleeding. Turkeys are gentle birds and if you are calm they will remain still. Try to approach jobs like this quietly, and don't stress out your animals.

REARING TURKEYS FOR MEAT

The methods for plucking and drawing turkeys are the same as for chickens, so see pages 34–37 as well as pages 88–89 for advice.

METHOD #20

TURKEY HOUSES

One of the key reasons that turkeys are popular to rear at home is that they are very large birds and grow very quickly. We always think big when we plan the housing for turkeys, as we know that their initial size is not a good gauge for what is to come at Christmas time. It doesn't have to be expensive, and the birds don't care about aesthetics - our shed was free and just needed a little patching up. And remember that turkeys like to perch, and also love space to wander around outside eating a range of seeds, insects and young green shoots.

A continuous supply of food and water

PLANNING A TURKEY HOUSE

A large shed can easily be turned into a turkey house. Even better to use, if available, is the inside of an old barn. Another option is to use an old chicken hutch, but it is essential that the hutch is thoroughly cleaned out with disinfectant before you start using it for turkeys.

Make sure that there are no major draughts in the turkey house, but that there is suitable ventilation. There is a thin line between the two, but you just need to make sure that your turkeys are protected from cold prevailing winds without living inside a sealed box. You will need a solid floor that can be covered with sawdust and straw. Alternatively, you can employ a deep-litter system (see opposite).

Other design elements to be aware of include a door that locks to prevent the wind from exposing your birds to hungry predators, and a ready supply of water and food.

POSITIONING A TURKEY HOUSE

Positioning a turkey house directly on the ground will lead to rising damp and can risk health problems for your birds. If you want to keep them healthy, put runners on the base of their flooring so that there is an air-flow underneath.

Never rear turkeys on the same patch of land 2 years in a row. (See page 82 for details of the parasite called Blackhead.)

BUILDING A TURKEY HOUSE

Once you have chosen a suitable shed or outbuilding to be turned into a turkey coop, you need to install large perches to the sides of the house where possible. These perches should be made of wood and should be at least 2.5–5cm (1–2 inches) thick, as mature turkeys can be very heavy.

FREE-RANGE

If you give your turkeys space to graze outside, they will taste better! It's a theory that makes sense because they are happier and healthier. Officially, free-range turkeys have continuous access to an outdoor area during daytime hours. The best space is one largely covered in mixed vegetation. Other benefits that result from their having more fresh air and sunlight include better eye and respiratory health. Turkeys that are given the opportunity to exercise and exhibit natural behaviour will also end up with stronger, healthier legs – ideal for big drumsticks.

FENCING

Fencing is essential to protect your turkeys from predators. Connecting an electric fence is an easy task and is a good skill to have when keeping poultry. Get hold of a fully charged deep-cycle or leisure battery and position it near the fence. Then take an energizer that will convert the power from the battery into a pulse of electricity to run through the fencing. These also come with an earth rod, which will need to be stuck into the ground. Use a watering can to dampen the ground so that there is a good connection for the first time. Your battery will last for a long time, as the power is very small, but when it comes time to recharge it why not consider a solar panel and a charge control unit? Cover the whole system with a large plastic or wooden box so that it is protected from rain and snow.

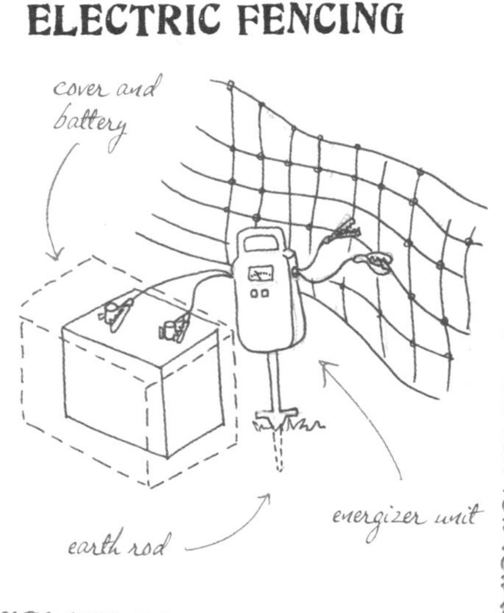

DEEP LITTER

This is a good method if you get it right. In effect you are creating a compost heap on the floor of the pen. Start with a 15cm (6 inch) layer of wood shavings. Then continue to mix pine shavings and straw with the high-nitrogen turkey poo. Keep adding enough shavings so that the floor has a good covering and the turkeys will do the mixing for you when they tread around or have a scratch. Do not remove any of the litter until the end of the year. This works best on a stone floor rather than a wooden base, as the moisture can damage the structure.

METHOD #21

PESTS & PROBLEMS

Turkeys are susceptible to disease and will sometimes just turn their toes up and die for no apparent reason. The good news is that there are things you can easily do to improve their chances – and prevention is always better than cure. The majority of people keep turkeys for a limited time before slaughter, and we find that in a home setting the slow-rearing breeds are slightly more hardy than the commercial types – plus, they require less food, so they are cheaper to rear.

BLACKHEAD

This parasite is the biggest threat to your turkeys – especially to young poults aged 3–18 weeks. Unfortunately, it is fairly common and can survive in the ground for at least a year – it is carried around in worms and snails. Blackhead is also the reason that chickens should be kept separately, as the parasite can survive in their gut without any obvious symptoms and they can pass it on to your turkeys.

Signs of a problem are sulphur-yellow droppings and more often than not a dead turkey. Your turkey will stop eating, look depressed and the wattle and comb on its head will turn a dark blue-black colour – hence blackhead! Treating poorly turkeys is difficult, so the best approach is prevention, keeping poultry separate and not moving turkeys on to land that has had chickens on it for 2 years after they've been removed.

PREDATORS

The main threats to your turkeys are foxes (see pages 26–27). However, young poults are also in danger of being killed by stouts, mink, rats and even hedgehogs. The best advice is to secure them with electric fencing and to put up a good solid turkey house that can be locked each evening. The electric fence doesn't keep turkeys in as much as acting as a barrier to keep foxes out. You need to make sure the poultry fence is tight and vertical, otherwise large birds can walk over it to escape, which will make them vulnerable.

Electric poultry fencing will allow for the movement of your flock and this movement of the turkeys around your plot will help with the health of your flock. If you intend to keep turkeys continuously such movements will protect from overgrazing and stop the build up of pathogens that may be shed into the soil.

It's best not to allow them to follow their instincts

FLYING AWAY

Turkeys like to roam. They stay together in social groups, but their woodland instincts mean that you will find them exploring an area in the region of 50m (150 feet) from their home. Despite being heavy birds, they can fly and will often be seen perched in the low branches of surrounding trees. We've never seen our turkeys taking off, yet in the evenings we've found them happily sitting on the roof of adjacent outbuildings. We learnt quickly and tend to put the turkeys to bed just prior to dusk as it saves the embarrassment of trying to find them in the evening gloom, never mind climbing on roofs in the dark. Our turkeys adopted a tendency to escape, which is why we started clipping their wings!

PROTECTING FLOCK HEALTH

Good flock health begins with obtaining eggs or stock from a reputable supplier with disease-free flocks. Most varieties are hardy and robust if basic precautions are taken. New birds need to be quarantined before introducing them to the rest of your flock. It is worth remembering that young poults are the most vulnerable as they wait for their immune systems to develop.

STRESS

When handling turkeys we are always very aware that stress plays a detrimental role in their wellbeing, so we are always as gentle and quiet as we can possibly be. It is relatively easy to see if your flock is suffering from stress such as overcrowding, dietary deficiencies (especially salt – which is in commercial feeds), insufficient food or watering, the birds will pluck their own feathers or the plumage of others, which may even escalate to cannibalism. The sight of blood induces pecking in turkeys so separate any injured birds until their wounds are healed and they no longer attract the attention of the others.

METHOD #22

LAYING & BREEDING

We tend to buy 6–8-week-old poults during the summer and rear them for the holidays, but you can hatch them yourself in spring if you are prepared to commit the extra resources. Hatching eggs yourself incurs extra costs over the spring period and requires more equipment for incubating, but it can make a really good business to run from home. Plus, you get to try turkey eggs, which, let's face it, are not something that everyone has eaten.

ESSENTIAL EQUIPMENT

- Nest boxes
- Water container
- Feeder
- Turkey breeding saddle
- Turkey breeding pellets
- Light and timer

EGG LAYING

Turkey eggs are not often available to buy, because the majority are used to hatch more turkeys for the next year. A hen will start laying from 28 weeks onwards, and will normally lay between late spring and mid summer. Obviously, the earlier eggs hatch into the turkeys that are the largest by the end of the year. If you want to encourage your hens to lay more, install a light in their pen that extends the sunlight hours to 14 hours a day.Keep it on a timer, and start lighting in early spring for an hour extra each week until you reach 14 hours a day. In a good season, that would span 16–20 weeks. Lighter breeds of turkey can lay up to 100 eggs; heavier birds lay about 50.

BREEDING

The starting point if you are attempting to rear your own turkeys is to select good breeding stock. You need to choose vigorous turkeys that have a uniform shape and colouring. The ideal ratio of hens to stags is 10:1, and they will be at their most fertile in their first year.

Change stags regularly so that inbreeding doesn't lead to genetic problems. It is worth removing hens that produce substandard poults. This type of selective culling can be difficult, but if you are trying to breed good turkey stock to sell then it is necessary. At home, breeding can be slightly less selective, but at the end of the day you want to get it right – turkey food is expensive and you will want to rear good table birds in return.

Some top tips for selecting a good breeding turkey are:

- The turkey should be in good health – no tendency to get ill.
- It should be walking well.

- It should have evenly distributed weight, with an upright carriage.
- It should be able to mate naturally.
- It should be a good egg layer with a high rate of hatchability.

Turkey breeds have been developed in such a way that natural breeding can be an issue. Some stags are too large to actually do the necessary to the hens so their semen is collected and hens are inseminated artificially. Several hens can be inseminated from each collection. Obviously, to get artificial insemination right you have to know what you are doing, so it is a task best left to the experts. You will probably not be surprised to know that it is not something we have had a desire to do ourselves.

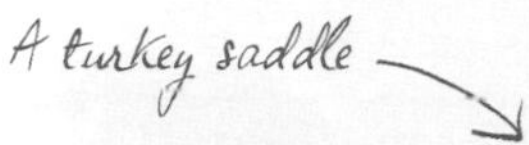

BROODY TURKEYS

A turkey will go broody like a chicken if you let it. However, turkeys don't make very good mothers, probably because their instincts have been bred out of them as a by-product of the massive commercial turkey business. A broody turkey hen will not want to leave her clutch and will sit for 25–28 days when they eventually hatch – there is always an exception to the rule, and she may rear the chicks just fine.

We have always wanted to try to breed turkeys, but to date we have listened to the couple who provide us with our poults who are convinced breeding is more effort than its worth. Having said that, with a life expectancy of up to 10 years keeping breeding stock is a long-term investment as they are too tough to roast. We particularly like the idea of a broody turkey – given how territorial they are and the loud cackling noise they can make, we imagine they must be ferocious and worthy of respect!

SADDLES

Late winter is the time of year when the placid stags start to puff up and pay much more attention to the squatting hens. It may sound funny, but putting a breeding saddle on the females will reduce any damage from him 'treading' her. The saddles, which are usually made of leather or canvas, sit on the females' backs and their wings pass through the straps, stopping the stags damaging them too much while mating. The saddles need to stay on in spring and come off in early autumn. The turkey may walk around awkwardly to begin with, but she will still be able to jump, fly and perform other turkey movements. During this period you should continue to check the hens for damage, and if there are any injuries, spray with iodine solution or powder.

METHOD #23

INCUBATING & HATCHING

The principles of incubating eggs are similar for all domestic poultry and fowl. Turkeys are relatively easy to hatch, but the key is that for the highest success rate you want to start incubation soon after the eggs are laid. Most people rear turkeys from young poults to provide the holiday birds for family and friends, so hatching and incubation isn't an issue for everyone. But if you do want to hatch your own turkeys, go ahead and give it a try.

ESSENTIAL EQUIPMENT

- Thermometer
- Spray bottle
- Incubator
- Torch
- Turkey crumb
- Incubating bulb – infra-red
- Brooder

FERTILIZED EGGS

Turkey eggs are equivalent in size to a duck egg and should be kept in cartons with the broad end upwards before you incubate them. Discard any cracked, damaged or misshapen eggs.

Store the eggs in a cool pantry or cupboard for no more than 1 week. For a few hours before they go into the incubator, allow them to acclimatize and reach room temperature. For every day you store eggs their chance of hatching decreases.

INCUBATOR

An automatic incubator is the simplest system for hatching turkey eggs and should produce the maximum yield. Remember that everything has to be clean. The eggs need to be thoroughly clean and the incubator also needs to have been cleaned with disinfectant. The temperature inside an incubator is exactly right for growing detrimental bacteria and micro-organisms, hence the need to sanitize the eggs and incubator.

Place the incubator in a room with a good even temperature and aim for an internal temperature of 37.5°C (99.5°F), with a humidity level of 55%, for the majority of the incubation period.

CANDLING

After the eggs have spent 1 week in the incubator, remove them and candle using a torch (see pages 30–31). Check that the embryo is fertile by looking for a veiny blob in the egg and make sure that the air sac is developing at one end. If the egg is clear, it is probably infertile. Red or black stains or a red blood 'ring' in the egg mean early death and any eggs that have speckling inside may have a bacterial infection and should be discarded.

HATCHING

After 25 days the eggs will start to hatch. To start with, there will be just a faint pip noise. This is the time to adjust the temperature to 37°C (99°F) and up the humidity to 75%.

By day 28 the turkeys should have broken free from their shells and will dry out into little puffballs.

When breeding turkeys, you should not count your turkeys before they are hatched. On average you can expect less than half the eggs to hatch and of the poults only about 30% can be expected to live to 2 weeks old.

BROODER

Once they have dried out, move them to a brooder with a heat lamp, chick crumb, water and sawdust bedding. Make sure they are checked regularly and that the brooder is protected from rats.

Brooders are usually circular or at the very least have curved corners, so the poults cannot crown into a corner and squash and suffocate each other. You will be able to see if the heat is right by observing the young birds. If they are cold they will congregate directly under the light where it is warmest. If they are too warm they will try and hide from the heat at the edges. As the birds grow the heat can be reduced by raising the heat lamps or reducing the voltage.

METHOD #24

SLAUGHTER

A turkey needn't be just for the holidays, but it is a good time of the year to slaughter and prepare the birds. Normally we slaughter and pluck 7 days before we intend to cook the birds. Like all poultry, turkeys should not be fed prior to slaughter, though actually it's probably more important here, as the birds can be hung to allow the flavour to develop. We hang ours in a cool outbuilding after killing and plucking.

PREPARATION

For us the holidays begin in earnest with the killing of the turkeys we have been keeping for the last 6 months. We usually have a mix of hens and stags, and part of the pleasure is seeing how much they have grown and what weight they have achieved. The first year we kept turkeys we were very excited by how big our flock had grown, but by the time they had been killed and plucked we realized that some were so large as to make them nearly impossible to fit into a domestic oven – not many families sit down around a 12.5kg (27.5lb) turkey.

Free-range turkeys, slowly reared, have a meat that has a very different texture from that of rapidly grown commercial birds. Their muscles have been used while the turkeys have been growing, so that when it comes time to slaughter you have a powerful bird to deal with.

It is essential that the bird has had no food in the proceeding 12 hours. As with all slaughter, it is important to be ready for the birds before you collect them from their shed/housing. The same system is used as for large ducks (see pages 54–55) or geese: an upturned traffic cone on a stool, broom handles to break the neck and a sharp knife to bleed the bird. To be efficient we find it necessary to have 3 of us involved in the killing.

SLAUGHTERING

Stag turkeys can be large and heavy, so they can be a handful, but they tend to be placid birds, so it's easy to pick one up and tuck it under your arm. When you turn it upside down to put it into the cone it will flap for a few seconds, then you need to fold the wings in as you lower it into the cone. It will be necessary to reach up under the cone to grasp the neck and pull the head through.

The broom handles may need to stretch further than for a duck, so you may need to raise the cone off the ground to allow for sufficient travel. As with the duck, you bend the head back and press down; however, you will need to put significant weight on the broom handles to break the neck. Bleed the bird and hang it up to pluck.

PLUCKING

Plucking a warm bird is easier than plucking a cold one, so it is best done immediately (see pages 34–35). You should pluck the body

all the way down to the little beard on the neck. As you would expect of a big bird, the wing feathers are hard work – time for the pliers. Plucking can take about 10–15 minutes per bird.

As we consider preparing the turkeys as the start of the festive season we tend to make the whole process, including plucking, a social event. Sometimes we have up to 4 people plucking a single bird, a couple of us tackling the wings (which take a lot of effort for very little reward in terms of meat) and 1 person plucking the front while the other plucks the back.

HANGING

The final act prior to hanging the plucked bird is to weigh it. Turkeys should be hung by their feet in a cool outbuilding. This hanging allows the flavour to mature and the flesh to reach perfection. We hang the birds for 5 to 7 days depending on the weather and tend to draw them on the day before we want to prepare them for cooking.

DRAWING

You should follow the same method for drawing a turkey as for a chicken (see pages 36–37). Drawing a turkey is much easier than drawing a smaller bird as there is plenty of room for you to get your hands inside.

We take great pride in presenting the final prepared bird with a set of giblets (heart, gizzard, liver and neck), neatly bagged and placed inside. At the end of this process, we also spend time removing any small quills that may have been missed when the bird was plucked.

A FESTIVE COUNTDOWN

- Slaughter and pluck your turkey 7 days before you intend to cook it.
- Hang the plucked bird for 5 to 7 days.
- Draw the bird the day before you intend to cook it and put it in the fridge.
- Roast your bird – then enjoy!

All trussed up and ready to go

5

RECIPES

INTRODUCTION TO

RECIPES

If you are keeping hens, eating the eggs is probably second nature and you should enjoy every very special mouthful. We've included a section on cooking eggs here because anyone keeping poultry will probably have a surplus at some stage and therefore being able to cook them well is pretty fundamental.

EATING EGGS & POULTRY

While many people keep poultry purely for egg production, the birds can of course be eaten as well. If you have gone to the trouble of rearing and keeping your own birds, you must surely ask yourself, why eat what you have become attached to? For us, the answer is because the quality of birds we rear ourselves is amazing. More importantly, we have had complete control of how they have been fed and looked after, including how they have been slaughtered and prepared. It's no fun killing your birds, but it is important to do it swiftly and humanely. After months of having cared for them, this is the final thing you do – so after that, do them the honour of using all their parts.

KEY SKILLS

To get the most from a bird, it is vital to use it all – some parts are rarely used and are vastly underrated. Therefore it is important to have a go at jointing a chicken as, along with boning, roasting and carving, it is one of the key skills to learn when it comes to cooking poultry. A roasted, stuffed bird is a meal that is all about sharing, and carving it at the table only adds to the pleasure. It becomes even more of a revelation if you bone it first, and why not stuff it with stuffing made to your own special recipe? The added advantage to boning the bird is that you don't have to be great at carving: all you need is a sharp knife and the ability to cut vertically (if you are at all concerned, buy an unsliced loaf of bread to practise on first).

It is probably safe to say we all realize it is wrong to waste our food, so having the confidence to use any leftovers and knowing that the meal will be every bit as tasty as the first incarnation is a very useful skill to acquire as well. Let's not forget that at the end, you will be left with a pot of stock that is crying out for you to make a broth, hopefully complete with pearl barley and split peas.

VARIETY IS THE SPICE OF LIFE

Different members of the family will probably have different favourite meals – for example, we love crispy duck, as well as the fact that confit melts in your mouth and keeps so well. Try different dishes, as you may discover that you have been missing out on a delicacy that will become your meal of choice. The delicious flavour and texture of goose liver definitely make it worth a try, while many people find that homemade chicken liver pâté is an easy introduction to cooking offal for the first time, with very popular results.

Eggs can be cooked in a huge variety of ways – here we're smoking the yolk

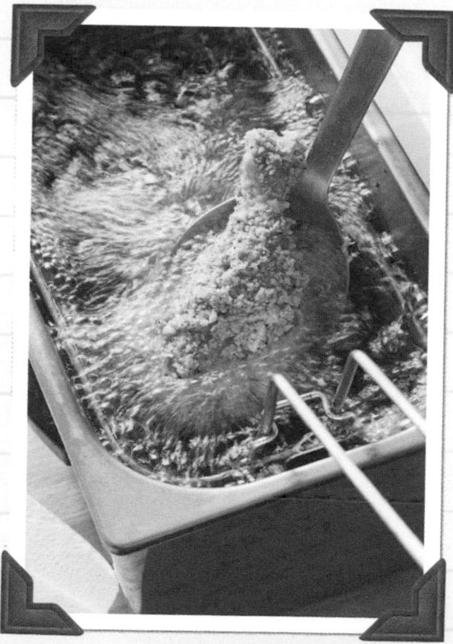

METHOD #25

COOKED EGGS

Your hens, ducks and geese will probably try and hide their eggs from you so be prepared to go hunting for them; it's worth it. Fresh eggs are delicious and there are many ways to cook them. Master these methods – and look after your hens – and you'll be able to whip up a delicious egg dish whenever you want. Eggs may all look very similar in shape, but there is a real difference between cooking hens' eggs and goose eggs.

STORING EGGS

Eggs should be stored in a cool, dry place and they will keep for about 3 weeks. Egg shells are porous, particularly those from ducks and geese, so it is best not to store them beside anything with a strong odour. Refrigeration isn't necessary, but if you have space in your fridge, great – however, make sure you bring them up to room temperature before cooking or you will have to cook them for longer to achieve the same results.

Eggs can also be frozen. Whole eggs are best lightly beaten and put into a freezer bag with all the air drawn out of it. Likewise, put egg whites into a freezer bag, squeeze out the air and seal the bag. If you freeze yolks they will need a pinch of salt or sugar beaten into them, otherwise they will become too thick to be usable.

TESTING FOR FRESHNESS

If you're using whole eggs that haven't been laid that day, you might want to test them for freshness. It is very simple, just put the egg in a bowl of water and if it floats it is not fresh.

BOILING EGGS

Size matters when it comes to boiling eggs, so all the guidance in the world will not on its own ensure that you have the perfect boiled egg – you will have to adjust the guidelines according to the size of your egg. Young hens (pullets) lay small eggs, older birds lay eggs that are 3 or 4 times larger. For us, the perfect boiled egg has the white cooked through and the outside layer of the

It is stale if it floats

BOILING TIMES

Times are for medium-sized eggs and start from when the water comes to the boil.

	HARDNESS	TIMING IN
Chicken	Soft	2½ minutes
	Medium	3 minutes
	Hard	4 minutes
Duck	Soft	2 minutes
	Medium	3½ minutes
	Hard	6 minutes
Goose	Soft	4 minutes
	Medium	6 minutes
	Hard	9 minutes
Quail	Medium	1 minute
	Hard	2 minutes

A proper breakfast

yolk has set. That is what we have based our timings on.

Remember to bring your eggs to room temperature and place the egg(s) in a small pan and just cover with salted water. Place on the heat and start the timings from when the water comes to the boil.

PERFECTING HARD-BOILED EGGS

The perfect cold, hard-boiled egg may sound like a no-brainer, but actually it is all too easy to spoil them, with the outside layer of the yolk discolouring as a result. So here's the foolproof way of boiling eggs:

Pop your eggs into the pan and just cover with salted water. Place on the heat and when the water has only just come to the boil take the pan off the heat, put a lid on and leave for 13 minutes. Then run cold water over the eggs until they are completely cold – a good 5 minutes. Peel the eggs and they will be perfect.

POACHING EGGS

First, choose your egg. The best eggs for poaching are the freshest ones because the whites stay together to form a perfect poached egg. The only eggs we wouldn't recommend poaching are goose eggs – this is because they are nearly all yolk, so they don't look very appetizing.

Use a saucepan (a shallow one will make it easier to get the eggs in and out) and fill it with salted water to a depth of about 5cm (2 inches). Add a couple of tablespoons of vinegar – we like to use cider vinegar. When the water is boiling, crack your egg into a cup and slide it into the water. Then turn the heat down and cook for your desired time. The perfect soft poached hen's egg takes 2–3 minutes, a duck egg about a minute more. Once they've cooked, take the eggs out of the water with a spoon.

Like fried eggs, it is also quite easy to see if the eggs are cooked properly – and remember, poached eggs should be runny inside.

CODDLING EGGS

Coddled eggs are whole eggs cracked into individual pots and cooked in a bain-marie. You can buy special lidded coddling pots made of porcelain, or you can use ramekins instead. Before you begin, place the coddling pots in a saucepan of boiling water (it should come about a third of the way up the pots). When they're heated up, add a knob of butter to each pot and when it has melted crack an egg into each one and season with salt and pepper. Put the lid on the coddler and then a lid on the pan – simmer for 8 minutes and the eggs are done.

There are lots of opportunities for adding flavour to your coddled eggs:

- Add a little finely chopped air-dried ham to the butter before adding the egg.
- Simply grate some nutmeg into the butter before you add the egg to the pot.
- Put a little finely chopped spinach into the butter for a couple of minutes before you add the egg.

SCRAMBLING EGGS

Scrambled eggs are simply eggs, butter and seasoning. The secret is to stop cooking the eggs while they are still very soft and to add a little extra butter at the end.

Beat the eggs well (for 4 people use 8 hen's eggs, 6 duck or 4 goose eggs – the ratio of yolk to white is much higher in duck and goose eggs, so the scrambled eggs will be much more golden in colour), adding a good pinch of salt. Melt about 50g (2oz) of butter in a non-stick pan, then pour in the beaten eggs and stir. When the eggs are cooked and still very soft, take them off the heat and add 25g (1oz) of butter. The eggs will continue to cook as the butter melts. Just before serving, add some black pepper.

Scrambled eggs lend themselves to extra flavours:

- Add a little smoked salmon. It doesn't take much – 50g (2oz) of chopped smoked salmon stirred into the eggs at the very end is enough for 4 people. Delicious served on toasted muffins.
- Try flavouring your eggs with very finely chopped parsley stalks, again adding them at the very end.
- For a light summery taste, skin, seed and chop a couple of tomatoes and add them at the last moment.

FRYING EGGS

Fried eggs are relatively foolproof if you use a non-stick pan and sufficient oil or butter. The amount of cooking is down to personal preference and will depend on how hard you like the yolk. For us, the perfect fried egg is soft and sunny-side up.

Heat some oil or butter in a non-stick pan over a medium heat. Crack the egg directly into the pan. Fry gently, allowing the heat to permeate through the egg, until the egg white is opaque and the yolk is your preferred consistency. Then remove with a fish slice, drain any excess fat and serve.

There's a good reason why the classic salads are still among the dishes we like to eat – they work! A hard-boiled egg on its own can be a slightly lacklustre ingredient, but in a Niçoise salad it becomes the star. Our version is very easy and quick to prepare.

SERVES 4

FOR THE SALAD

400g (13oz) new potatoes

125g (4oz) French beans, topped and tailed

4 eggs, hard-boiled, cooled and peeled

200g (7oz) tomatoes, chopped

100g (3½ oz) pitted black olives

150–250g tinned tuna

1 tin of anchovies (approx. 12 fillets)

2 red onions, sliced (optional)

a couple of handfuls of mixed salad leaves

FOR THE VINAIGRETTE

100ml (3½ fl oz) olive oil

2 tablespoons white wine vinegar

1 tablespoon chopped fresh chives

1 tablespoon chopped fresh parsley

1 teaspoon lemon zest

salt and freshly ground black pepper

SALAD NIÇOISE

Bring a pan of salted water to the boil. Cut your new potatoes in half, add them to the pan and cook for 15 minutes, or until tender. At the same time, bring a smaller pan of salted water to the boil, add your green beans and blanch for a few minutes. Drain the potatoes and beans when ready, then leave to cool.

Cut the hard-boiled eggs in half. Now assemble the tomatoes, olives, tuna and eggs in a bowl with the salad leaves, then add the cooled potatoes and beans, the anchovies and the onions, if using.

Make the vinaigrette by mixing all the ingredients in a bowl, seasoning with salt and pepper. Stir with a small whisk and drizzle over the salad just before serving.

Duck eggs are richer than hens' eggs, and this Scotch egg makes a hearty snack. It goes well with chutney or a sharp sauce. For this recipe, the fattier the duck meat the better. If it is lean, scrape the duck fat from any leftover skin, or use about 1 tablespoon of the partially solidified fat from the roasting tray.

SERVES 4

400g (13oz) cooked duck meat
juice of ½ an orange
zest of 1 orange
3 spring onions, finely sliced
4 duck eggs, soft-boiled, cooled and peeled
3 tablespoons plain flour
salt and freshly ground black pepper
1 egg, beaten
100g (3½ oz) fresh breadcrumbs
vegetable oil, for deep-frying

DUCK SCOTCH EGGS

Put the duck meat, orange juice and any extra duck fat (if using) into a food processor and whizz for about 60 seconds, or until a ball of chopped meat starts to form. Transfer to a bowl, add the orange zest and spring onions, and mix thoroughly. Divide the mixture into 4 balls and flatten each one out on a lightly floured surface until big enough to wrap around a duck egg.

Put the flour, seasoned with salt and pepper, on to a plate, and the beaten egg and breadcrumbs into 2 bowls. Roll the meat-covered duck eggs in the flour, then dip them into the beaten egg and finally roll them in the breadcrumbs.

Heat some vegetable oil to 185°C (365°F) in a deep-fryer or a large pan and deep-fry the Scotch eggs for 5 minutes, or until they are a deep golden colour. Drain the eggs on kitchen paper.

Duck eggs wrapped in a duck-meat blanket

Our version of this classic dish is quick and easy. Once made, the hollandaise sauce cannot be reheated, so this is all about timing. You can pop the blender jug into hot water to keep the sauce warm, but you will need to keep stirring it.

SERVES 4

12 rashers of dry-cured bacon (smoked bacon is also very tasty)

4 eggs

cider vinegar, for poaching

2 English muffins, split in half

FOR THE HOLLANDAISE SAUCE

200g (7oz) butter

3 egg yolks

½ teaspoon Dijon mustard

salt and white pepper

EGGS BENEDICT

Cook the bacon in a dry frying pan until crisp, then set aside to keep warm. Put a pan of water on to boil for the poached eggs, with a dash of vinegar and a pinch of salt (see page 96). Heat the butter in a small pan. Place the muffins in the toaster ready to go. Put the egg yolks and mustard into the blender.

Put the eggs into the boiling water and turn down the heat.

Turn the toaster on – the 2 minutes for the muffins is a good timer for the poached eggs – turn the heat off under the pan the moment the toast pops up.

Turn the blender on – if it has been a little violent and thrown yolk up the glass, scrape it down into the blender again. After 30 seconds, start dribbling in the warm butter – it should take you about a minute.

Time to assemble: muffin, bacon, poached egg, hollandaise (don't worry about buttering the muffin, as there is enough butter already in this dish).

This pizza was designed with poached eggs in mind. The crispy crust, melted cheese and soft spinach all work well together to make it a pleasure to eat. Remember to keep your egg slightly runny so that it oozes all over the pizza.

SERVES 2

FOR THE SAUCE

1 tablespoon olive oil

1 garlic clove, chopped

50ml (2fl oz) passata

1 tablespoon chopped fresh basil

salt and freshly ground black pepper

FOR THE PIZZA BASE

150g (5oz) plain flour

2 tablespoons olive oil

25–50ml (1–2fl oz) cold water

FOR THE TOPPING

150g (5oz) spinach

75g (3oz) black olives, pitted

50g (2oz) mozzarella cheese

50g (2oz) Gorgonzola cheese

1 teaspoon cider vinegar

1 large duck egg

PIZZA À LA DUCK EGG

Preheat the oven to 220°C (425°F), Gas Mark 7.

To make the sauce, heat the olive oil in a pan, then add the garlic and cook for a couple of minutes until softened but not coloured. Add the passata and the basil, stir, then season with salt and pepper. Set aside to cool.

To make the pizza base, put the flour and oil into a food processor or mixing bowl and whizz or stir together, adding enough water to make a workable dough. Roll the dough out on a floured surface until 5mm (¼ inch) thick and spread the pizza sauce mixture over it.

Bring a large pan of water to the boil and blanch the spinach briefly, then squeeze the leaves to remove as much water as possible. Spread the leaves over the pizza and add the olives, mozzarella and Gorgonzola. Season with salt and pepper and place in the oven for 8 minutes.

While the pizza is cooking, boil a small amount of water in a shallow pan and add the cider vinegar. Carefully crack in the duck egg and poach for 3–4 minutes, or until cooked, then remove it from the pan and drain on kitchen paper. Take the pizza out of the oven and place the poached egg in the middle. Put back into the oven for a further 3 minutes. Serve straight away.

This is a show-stopper for a vegetarian brunch, but if you prefer you can swap the halloumi for a couple of pieces of smoked streaky bacon. For the smoking all you'll need is foil, rice and tea - and trust us, it's worth having a go.

SERVES 4

1 tablespoon sugar

1 tablespoon loose tea leaves

1 tablespoon uncooked rice

4 eggs

2 slices of bread

1 tablespoon olive oil, plus extra for frying

salt and freshly ground black pepper

1 tablespoon orange zest

4 pickled baby beetroots, or 2 large ones

1 tablespoon balsamic vinegar

4 slices of halloumi cheese

a pinch of paprika

salad leaves

SMOKED EGGS WITH HALLOUMI

Make a liner of foil for a deep pan (the foil liner should be large enough to come well above the sides). Mix together the sugar, tea leaves and rice in a small bowl and place in the centre of the foil liner. Put the lid on and seal it by folding up the edges of the foil so that the smoke is contained. Put the pan on a high heat for 2 minutes.

Next separate the eggs (the whites are not used in this recipe, so freeze them) and place the yolks in individual baskets of lightly oiled foil. Make these by cutting 10cm (4 inch) squares and moulding them into a simple bowl shape. Take the lid off the pan, set the egg baskets on top of the smoking mixture and then replace the lid. Let the eggs smoke over the heat for a further 3–4 minutes.

To make the croutons, cut the bread into thin strips and drizzle them with the olive oil. Sprinkle them with salt, pepper and the orange zest. Fry in a hot pan until crispy, then remove from the pan and leave them to drain on kitchen paper.

Cut the pickled beetroot into small matchstick strips and put them into a bowl. Add the balsamic vinegar and leave them to soak for 5 minutes. While the croutons are draining, use the same frying pan and another drizzle of oil to cook the slices of halloumi, adding a pinch of paprika for an extra bit of pizazz. Season with salt and pepper and serve on top of the salad leaves and croutons, with the smoked eggs and pickled beetroot.

These tarts smell great as they come out of the oven, but they are best when warm, not hot. So if time allows, leave them to stand for half an hour before eating. You can obviously fill the tarts with any flavours that take your fancy, but we love this combination of sweet onion and tart goats' cheese.

SERVES 4

FOR THE PASTRY

150g (5oz) plain flour

75g (3oz) cold butter, cut into small cubes

1 tablespoon cold water

FOR THE FILLING

1 red onion, peeled

200g (7oz) soft goats' cheese

2 eggs

1 tablespoon fresh thyme leaves

200ml (7fl oz) cream

salt and freshly ground black pepper

CARAMELIZED ONION & GOATS' CHEESE TART

Preheat the oven to 180°C (350°F), Gas Mark 4. Sift the flour into a bowl, then add the butter and with your fingertips gently rub it into the flour until you have something resembling fine breadcrumbs. Add the water and gently pull the dough together into a ball.

Roll out the pastry into a thin circle on a lightly floured surface and gently press it into a 20cm (8 inch) flan tin with a removable base. Prick the pastry with a fork every 1cm (½ inch) or so. Cut a circle of greaseproof paper to fit the base, then crumple it up and open it out again so that it lies flat in the base. Put a layer of baking beans or rice on top of the paper and bake the base in the oven for 15 minutes. Remove the beans, then bake blind for a further 5 minutes. Set aside to cool.

When you're ready to make the tart, reheat the oven to 180°C (350°F), Gas Mark 4, if necessary. To make the filling, slice the onion into segments. The layers of each segment should still be held together at the base. Heat a frying pan until it's very hot, then put the onion segments on the dry pan and leave until caramelized, then turn over and caramelize the second side. This will only take a couple of minutes.

Arrange the onion segments on the tart base, then arrange two-thirds of the goat's cheese in large pieces over the onions. Beat the eggs well, then add the remaining goat's cheese, all but a few thyme leaves, the cream and a little salt and pepper and beat for a further 15 seconds.

Carefully pour the egg mixture into the tart case and bake in the oven for 25 minutes, or until golden brown. Just before serving, sprinkle with the rest of the thyme leaves.

With a little green salad this makes a lovely lunch. The principles are very easy, and you can adapt the recipe to use anything you like instead of, or as well as, potatoes and onions.

SERVES 4

extra virgin olive oil

1 onion, halved and sliced

3 garlic cloves, sliced

4 medium potatoes, peeled, halved and cut into 5mm (1/4 inch) slices

8 eggs

salt and freshly ground black pepper

grated cheese (optional)

FRITTATA

Pour a good glug of extra virgin olive oil into a shallow non-stick frying pan over a medium heat. Add the onion and garlic and cook gently for a few minutes, then add the potatoes and cook until they start to soften (about 10 minutes) – you don't want the onions and potatoes to colour at this stage, so stir/toss them regularly.

Beat the eggs in a bowl and add some salt. Pour the beaten eggs into the pan and after about 20 seconds use a spatula to pull the cooked egg around the edges into the middle – the uncooked liquid egg will flow out to replace it (this will make your frittata lighter). Make sure you keep the potatoes and onions evenly distributed.

Having 'aerated' your frittata, let it cook from the bottom up. At this stage traditionalists would explain how to turn the frittata by sliding it on to a plate and flipping it over; however, the easiest way to cook the top is to pop it under a preheated hot grill for a minute. If you are including grated cheese in your frittata, sprinkle it on top before popping it under the grill.

When the frittata is cooked, slide it on to a plate and crack some pepper over it. For the best flavour, let it cool down a little and serve it warm or at room temperature, rather than hot or straight from the fridge. Serve cut into wedges.

Duck eggs are great for baking – it must be either the richness of the yolk or the ratio of yolk to white. Either way they produce successful cakes. Be aware that duck eggs vary in size: our Indian Runners' eggs are the same size as a large hen's egg, whereas our Muscovy ducks produce eggs that are nearly twice the size.

SERVES 6–8

- 100g (3½ oz) softened butter or margarine
- 125g (4oz) self-raising flour
- 1 teaspoon baking powder
- 100g (3½ oz) caster sugar, plus extra to decorate
- 2 duck eggs, beaten
- 150ml (¼ pint) cream, whipped
- 2 tablespoons strawberry jam

DUCK EGG VICTORIA SPONGE

Preheat the oven to 160°C (325°F), Gas Mark 3. Lightly butter two 18cm (7 inch) cake tins and put a circle of greaseproof paper on each base.

Hold your sieve up high and allow the flour and baking powder to fall through it into a large bowl. Cut the butter into cubes about 5mm (¼ inch) across and add them, along with the sugar and the eggs, to the flour. Whisk until thoroughly combined. The mixture should be loose enough to fall easily off a spatula. If it appears to be too thick, add a teaspoon of warm water and whisk again.

Divide the mixture between the prepared tins and bake in the oven for 30 minutes. When cooked, remove from the oven and cool on a baking rack.

Once cool, spead a layer of whipped cream on the top of one of the sponges and a layer of strawberry jam on the bottom of the other. Gently sandwich them together and sprinkle caster sugar over the top.

METHOD #26

ROAST BIRDS

There is something very impressive about serving a whole roast bird. The preparation can take a little time, but while the bird is in the oven, and during the resting time after it comes out, you will be able to sort out the accompaniments. For larger birds the oven can be on for hours, so it's worth considering how you can use any spare oven space. A tray of vegetables put in the oven for the last 45 minutes of the cooking time and then left to finish roasting while the bird rests is an easy way of making the most of your oven and your time in the kitchen.

PREPARING TO ROAST

Many factors will have an influence on cooking temperatures and times. Our advice will help you to cook a perfect roast bird, but be prepared to adjust the directions based on knowledge of your own oven.

First of all, you need to know your oven. They are all different, so you may need to turn your bird around to ensure even colouring. In addition, a bird taken from a cold fridge and popped straight into the oven is very different from one that has been allowed to come up to room temperature. It's not wise to leave a bird on a work surface for hours prior to cooking, so remove it from the fridge half an hour before cooking. It is also worth noting that the air-flow through a stuffed bird is very different from one with open body cavities. This means that cooking times may vary. To ensure your bird is cooked, always test for doneness (see page 117).

SELECTING A TRAY

The only equipment you really need is a good tray. A roasting tray is the most usual way of roasting; however, we have an old-fashioned self-basting roasting tin with a lid. Steam collects on the dimples on the lid and drips on to the bird, basting it throughout the cooking process.

Your roasting tray should not be too much larger than the bird you are roasting (so the juices don't spread out and evaporate, causing the residue to burn). If you are roasting duck or goose, place a trivet in the tray to keep the bottom of the bird out of the fat and allow it to roast rather than deep-fry.

UNDERSTANDING THE PRINCIPLES

The following basic principles can be followed for roasting birds.

- Weigh your bird and calculate the temperature and timings required.
- Preheat your oven.
- Choose a roasting tray that is the right size for your bird.
- Halfway through the cooking time check on the bird's progress, so that you can protect it with foil if it is browning too fast, or baste it if necessary.
- Protect the breast meat of poultry during roasting to stop it becoming dry.

- When roasting time is complete, test the bird for 'doneness'.
- Once roasted, allow the bird to rest in a warm place.

STUFFING YOUR CHICKEN

We usually place a quarter of a lemon, or orange, and a couple of garlic cloves inside the body of the bird to add flavour. If you don't want to stuff the chicken you can simply put soft butter, or a mix of butter and your favourite herbs (we love tarragon), under the skin before you roast it.

A small chicken will benefit from stuffing, both to keep it moist and to make it go further. Start by pushing your fingers under the skin of the breast to loosen it, then push the stuffing under the skin. Cover about half the breast, then use the remaining stuffing to fill up the large flap of skin. Turn the bird over and use a skewer to hold the flap of skin in place. Rub the bird with butter and roast for the required amount of time (see pages 116–117 for more details).

An alternative method is to smear the bird with butter, put about 300ml (½ pint) of water into the roasting tray along with some carrots, leeks and onions, then roast it. This 'French Roast' provides a very tasty liquor for the gravy, but the bottom of the bird is not crisp as with conventional roasting.

KEEPING THE CHICKEN MOIST

The breast meat of most poultry has to be protected during roasting to stop it becoming dry. This can be done by pushing any forcemeat stuffing up under the skin (so the fatty sausage meat slowly bastes the breast), by placing streaky bacon on the breast for the first part of the cooking, removing it after basting (it will be lovely and crispy and is definitely a 'chef's perk'), or just by regular basting.

A SIMPLE STUFFING FOR CHICKEN

For a 2.25kg (5lb) chicken

500g (1lb) sausage meat
2 rashers of streaky bacon, chopped
the liver from the giblets
zest of 1 lemon
150g (5oz) fresh breadcrumbs
3 tablespoons chopped fresh parsley
salt and freshly ground black pepper

Mix all the stuffing ingredients together in a bowl.

HOW TO STUFF A CHICKEN

Gently push your fingers up under the skin of the breast from the neck end. Push the stuffing in so it covers about half the breast and fill the large flap of skin with the remainder.

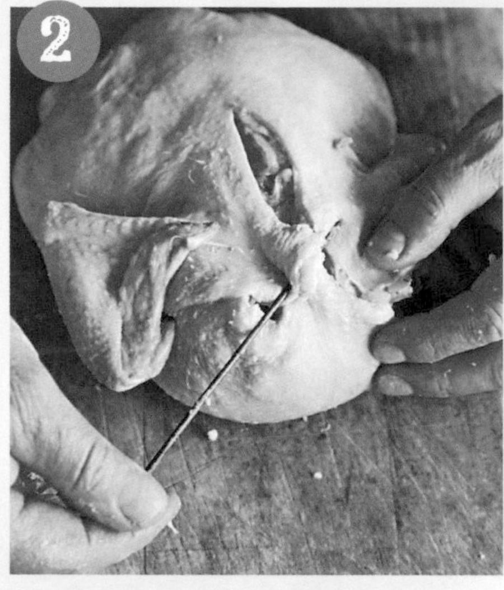

Turn the bird over and use a skewer to hold the flap of skin to the base of the bird.

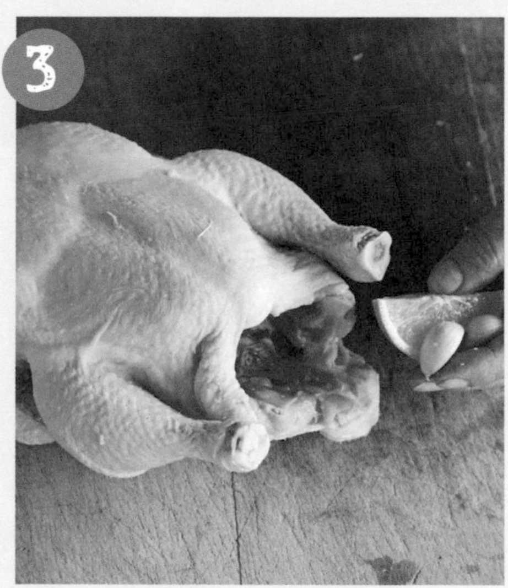

Place a quarter of an orange and a couple of garlic cloves into the body of the bird to add flavour.

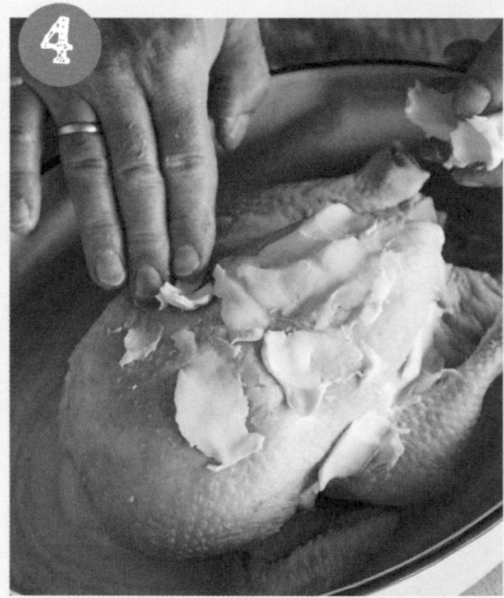

Finally, smear the bird with butter and roast for the necessary amount of time.

ROASTING YOUR CHICKEN

Preheat the oven to 200°C (400°F), Gas Mark 6, and calculate your cooking time, allowing 15 minutes per 500g (1lb). When judging cooking time, always calculate it using the combined weight of bird and stuffing (if used). Halfway through the cooking time, check on the bird's progress. If the skin seems to be getting too brown, loosely place some foil over it for the latter stages of cooking. If necessary, turn your bird around to ensure even colouring as it roasts.

Regular basting will help to keep your bird moist. To baste, simply tip the pan to one side and spoon any fats over the bird. The disadvantage of lots of basting is that heat can escape from the oven while you are doing it, so always take the bird out of the oven, close the door and place the chicken on top of the stove to baste. You should baste the bird every 20 minutes or so.

TESTING FOR DONENESS

When the roasting time is complete, test the bird for 'doneness'. There are 3 ways to test for doneness:

- USING A MEAT THERMOMETER – When placing your thermometer, ensure that you do this in the thickest part of the bird, either in the breast above the wing bone, or through the thigh. It should not be in contact with any bone and should read at least 70°C (158°F) when the bird leaves the oven.
- USING A SKEWER – If you push a skewer into the thickest part of the breast and the thigh, the juices that run out when the skewer is removed should be clear, with no trace of pink.
- TUGGING THE LEG – If you gently pull the leg of a cooked turkey or chicken away from the body, it should 'give' rather than dragging the bird with it.

RESTING THE CHICKEN

Once the bird is cooked, remove it from the oven, cover it loosely with foil and allow it to rest for 10 minutes in a warm place. Resting makes the whole carving process easier, as the fibres of the meat relax and the juices are spread evenly through the bird.

MAKING GRAVY

No roast is complete without gravy made from the fats and juices from the pan, so it is important not to waste all this extra flavour. Make the gravy in the roasting tray itself to ensure that no flavour can escape. Pour all the juices into a heat-resistant jug and after a couple of moments the fat will have separated and formed a layer on the top. Pour 3 or 4 tablespoons of fat back into the tray and store the rest in the fridge. Roast duck and goose produce lots of fat and it is great for

HOW TO CARVE A ROAST CHICKEN

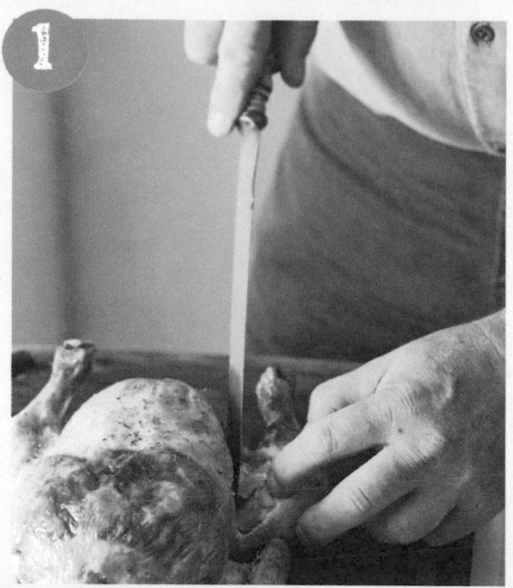

Start by removing the legs. Pull the leg away from the body and slip a knife through the joint to cut it off. Repeat on the other side.

Divide the legs into drumstick and thigh. You can slice meat off the joints at this stage

Cut slices from the stuffing end of the breast. The thickness of the slices is down to personal choice, but warm meat is better sliced slightly thicker (about 5–8mm/¼–½ inch thick).

Place the slices of meat on warmed plates, as sliced breast meat can cool very quickly, and serve immediately.

roasting potatoes or for frying with. Spread a tablespoon of flour in the tray to absorb the fat and use a spatula to scrape up all the flavour. Place the tray on a burner and gently heat to cook the flour. Add a glass of wine, if you like, whisk it in and cook it for a minute, then add the juices. If you have insufficient juices to make your gravy, add some of the water that you used to cook your vegetables or some stock. You might also like to try whisking a tablespoon of cranberry jelly into the gravy just prior to serving.

CARVING & SERVING

First, remove the legs. Hold 1 drumstick firmly with your fingers, pull gently away from the bird, then cut through the skin and hip joint to separate the leg. Cut away the 'oyster' (2 small pieces of meat located on each side of the bird's backbone) and separate the drumstick and thigh. Cut slices of drumstick and thigh meat parallel to the bone.

Next, remove the wings by making a cut deep into the breast to the body frame, parallel to and as close to the wing as possible. The closer you get, the more slices of breast you will get.

Starting at the front, where the breast is plumpest, cut thin slices of white meat – the initial cut will be mainly skin, but thereafter the slices get bigger. The slices should fall away from the bird as they are cut. Continue carving until there is enough meat for your guests and serve with gravy and all the trimmings.

NOW TRY: ROAST TURKEY

Most people have their own 'family secret' approach to cooking the perfect roast turkey, as they tend to be eaten only once or twice a year. We stuff our turkey using a stuffing similar to the chicken recipe on page 115; however, we do sometimes vary it by adding roughly chopped chestnuts and apricots. For a turkey, you will need about 3 times the quantity of stuffing you would make for a chicken. As the birds are bigger, the breast meat needs to be protected. Cover the turkey with a layer of streaky bacon on top of the smeared-on butter, then loosely place foil over it until the last stages of cooking, when it can be allowed to brown.

Smaller birds, up to 6kg (12lb), are best cooked like chicken on a higher heat for a shorter period of time. Preheat the oven to 190°C (375°F), Gas Mark 5, and allow about 10 minutes per 500g (1lb). For larger birds, preheat the oven to 160°C (325°F), Gas Mark 3, and allow 20 minutes per 500g (1lb). Baste several times during the cooking and rest before serving.

We rest our turkey (which tends to be 7–9kg/14–18lb in size) for the whole duration of cooking the roast potatoes in the oven (approximately 45 minutes).

NOW TRY: ROAST DUCK

Preheat the oven to 220° (425°F), Gas Mark 7. Trim a duck of any excess fat and remove the wing tips and the parson's nose (the fleshy part of the tail). Ensure that the cavity opening is clear. Place the duck on a rack over the sink and slowly pour a kettle of boiling water over it. Pat the duck dry and, while it's still warm, prick the skin all over. Slice an orange into 5mm (¼ inch) thick slices with the skin on and place the slices in the cavity with a couple of garlic cloves. Put the duck, breast side down, on a trivet in a roasting tray. Turn the oven down to 180°C (350°F), Gas Mark 4 and place the duck in the oven. Allow about 20 minutes per 500g (1lb). Remove from the oven, turn the duck breast side up and return to the over for a final 15 minutes to crisp. After roasting, the orange and garlic can be taken out and made into a sauce using the stock made from the giblets.

NOW TRY: ROAST GOOSE

As with duck, any fruity stuffing will go well with goose, or try putting an onion, some herbs and some garlic into the body cavity prior to roasting. Don't throw away the giblets. Put them in a pan with a roughly diced onion, 2 carrots and 1.5 litres (2½ pints) water and simmer, uncovered, for the duration of the roasting. You can use this stock to make a gravy or sauce to serve with the goose.

Always set the bird on a trivet in the roasting tray in order to keep the bird out of the oil – you want the bird to roast not fry.

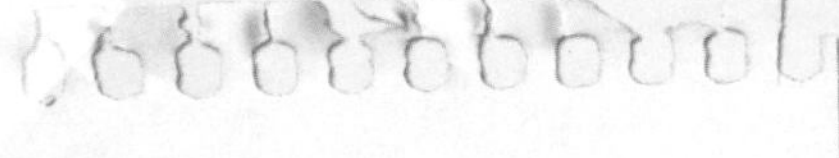

HOW TO ROAST A GOOSE

For a 5kg (10lb) goose

- salt and freshly ground black pepper
- 1 onion
- sprigs of fresh thyme
- sprigs of fresh sage
- 2 teaspoons oil

Preheat the oven to 220°C (425°F), Gas Mark 7, then turn down to 200°C (400°F), Gas Mark 6, when you put the goose in the oven. Follow the instructions opposite to roast your goose.

Prepare your goose by removing all the fat from inside the cavity.

Rub the breast and legs with oil and season generously with salt. Sit the bird, breast side up, on a trivet in a roasting tray.

Use a skewer to prick the skin, especially under the wings and around the tail.

Season the inside of the bird and stuff it with the onion and herbs.

Cover with a piece of foil. Place the goose in the oven for 1½ hours. Then take it out and remove the foil. Pour the fat out of the tin and use it to lightly baste the goose. Re-cover with foil and roast for another 1½ hours, then baste as before.

Return the goose to the oven without any foil to brown for a final half-hour, then transfer it to a large board or platter to rest in a warmish place for 30 minutes before serving.

After a roast dinner there are nearly always leftovers and these can make another meal that is just as tasty. In fact, we regularly cook more vegetables than necessary so that we can have bubble and squeak. It's always served with the reheated leftover gravy, slices of cold meat and homemade chutney.

SERVES 4

1kg (2lb) mixture of cooked vegetables, e.g. mashed/boiled/roast potatoes, boiled/roast parsnips, broccoli, cauliflower, Brussels sprouts, carrots, peas, leeks, cabbage, swede, turnips

dripping or sunflower oil

TO SERVE

Cooked turkey meat

Gravy

Chutney

TURKEY WITH BUBBLE & SQUEAK

Cut all your leftover vegetables into small chunks.

Put all the vegetables into a large bowl. If the mixture looks as though it will fall apart, use your hands to squash some of the larger pieces so that they bind everything together. Form the mixture into patties about the size of a large burger – if you like your patties to be perfectly round, you can trim them into shape them with metal rings.

Heat a few tablespoons of dripping or sunflower oil in a frying pan and add the patties to the hot fat. Fry for a few minutes to allow a crust to form, then turn them over. Reduce the heat and leave them to cook for a further few minutes more, until they are heated through.

Serve the bubble and squeak with cold roast turkey, hot gravy and some homemade chutney.

Mix the vegetables well

Turkey pie is another classic leftover meal that makes the most out of both the white and the brown meat. Turkey, like chicken, which can easily be substituted here, goes incredibly well with tarragon. Fresh tarragon will give the best flavour, but the freeze-dried variety will also do the job.

SERVES 4

50g (2oz) butter

2 garlic cloves, finely chopped

1 onion, finely chopped

150g (5oz) mushrooms, sliced

1 tablespoon chopped fresh tarragon

75ml (3fl oz) white wine

400g (13oz) cooked white and brown turkey meat, chopped

400ml (14fl oz) milk

1 bay leaf

2–4 tablespoons plain flour

salt and freshly ground black pepper

400g (13oz) ready-made puff pastry

1 egg, beaten with 1 tablespoon milk

TO SERVE

potatoes

broccoli

TURKEY PIE

Preheat the oven to 180°C (350°F), Gas Mark 4.

Melt the butter in a large pan and add the garlic and onion. Cook gently until softened, then add the mushrooms, tarragon, wine and turkey meat. Cook for 5 minutes.

In another pan warm the milk with the bay leaf. Gently whisk in the flour a bit at a time until the sauce starts to thicken. When it reaches a good consistency and there are no lumps, add it to the pan of turkey. Stir and simmer for 2–3 minutes, then season well with salt and pepper and pour into individual pie dishes. Set aside to cool a little while you get the pastry ready.

Roll out the puff pastry to less than 1cm (½ inch) thick, cut out 4 circles and lay on top of the dishes (you can use the leftover pastry to make leaf decorations for the crust, if you like). Make a few holes in the top so that the filling doesn't overflow, brush the pastry with the beaten egg and bake in the oven for 20 minutes until the pastry is golden and puffed up– more than enough time to cook some potatoes, steam some broccoli and wash up.

This Thai-style curry is very easy to make. The aim should be to produce a tasty dish rather than a particularly hot one – getting the balance right here involves choosing chillies with a good flavour, not just the really hot ones.

SERVES 4-8

- 2 tablespoons vegetable oil
- 1 butternut squash, roughly chopped
- salt and freshly ground black pepper
- 500g (1lb) fresh turkey meat, cut into 2.5cm (1 inch) cubes
- 1 tablespoon Thai green curry paste
- juice of 3 limes
- 2-3 fresh red chillies, chopped
- 2 teaspoons nam pla (fish sauce)
- 1 teaspoon coriander seeds, crushed
- 4 tablespoons chopped fresh coriander
- 2 teaspoons lemon grass paste, or 1 stick of fresh lemon grass, chopped
- 2 x 400ml (13 fl oz) tins of coconut milk
- 2 tablespoons sesame oil
- 6-8 spring onions, chopped
- fresh coriander leaves, to garnish

TURKEY CURRY

Heat the vegetable oil in a large pan and add the squash. Season with salt and pepper and fry for a few minutes, then add the turkey, curry paste, lime juice, chillies, fish sauce and crushed coriander seeds. Cook for 5–10 minutes, then stir in half the fresh coriander and the lemon grass. Pour in the coconut milk, then reduce the heat and simmer gently for 20 minutes, or until the squash is tender.

Stir in the sesame oil, then add the chopped spring onions and the rest of the chopped coriander. Serve with plain rice and a final sprinkling of fresh coriander leaves.

This recipe is a great way of making the most of turkey leftovers – it's delicious served with homemade redcurrant jelly. We love this combination of flavours, but feel free to adjust the spices to suit your own taste.

SERVES 4

- 400g (13oz) cooked white and brown turkey meat, shredded
- 1 teaspoon wholegrain mustard
- 1 teaspoon Dijon mustard
- a small pinch of ground cloves
- zest and juice of 1 orange
- 2 teaspoons finely chopped fresh thyme leaves
- a pinch of grated nutmeg
- 75g (3oz) softened butter
- salt and freshly ground black pepper
- a knob of butter, melted

POTTED TURKEY

Put all the ingredients except the melted butter into a blender and blitz for a few moments, using the pulse control to keep the texture rustic, rather than very smooth. If you don't have a blender you can put everything into a large bowl and mix with a fork.

Divide the mixture between 4 ramekins and press firmly down, then cover with a layer of melted butter to seal. Leave in the fridge to set. The butter will set very quickly in the fridge and the potted turkey can be kept for up to a week.

Serve the potted turkey with redcurrant jelly and slices of toast.

A roast goose is a seriously large meal, so this warm salad is a lovely way to give leftovers a new lease of life. Red gooseberries are sweeter than their green counterparts and work well as a sauce here, but green ones will do fine.

SERVES 2-4

FOR THE SALAD

300g (10oz) cooked goose meat

2 heads of chicory

2 little gem lettuces

olive oil

salt and freshly ground black pepper

4 spring onions, sliced

50g (2oz) walnuts, chopped

2 handfuls of pea shoots

FOR THE SAUCE

250g (8oz) gooseberries (red ones if possible)

2 tablespoons port

50g (2oz) sugar

75ml (3 fl oz) water

1 tablespoon orange zest

GOOSE SALAD WITH GOOSEBERRIES

To make the sauce, cut the gooseberries in half and put them into a pan with the port and sugar. Place over a medium heat for 10 minutes. Once the alcohol has burned off and the gooseberries start to soften, add the water and orange zest. Cook for a further 5–10 minutes, then strain the sauce through a fine sieve and keep warm.

Cut the goose meat into chunks and slice the chicory and lettuce into thin strips. Heat a little olive oil in a pan, add the goose and sliced leaves and toss quickly, seasoning with salt and pepper. Add the spring onions, walnuts and pea shoots and heat for another couple of minutes.

Serve the goose salad with the warm gooseberry sauce drizzled over the top.

Lovely little leaves

Braising involves browning the meat before cooking it slowly in added liquid. It's a fantastic method of cooking duck that ends up with a delicious, full-flavoured sauce, and the meat is beautiful, tender and moist.

SERVES 4

- 1 x 2kg (4lb) duck
- 450g (14½ oz) smoked streaky bacon
- 1 carrot
- 1 onion
- 150g (5oz) chestnuts
- 1 bouquet garni, containing sprigs of fresh thyme, a bay leaf, a fresh sage leaf and peppercorns
- 125ml (4fl oz) white wine
- 300ml (½ pint) chicken stock
- a knob of butter
- 8 chipolata sausages
- 12 shallots, peeled
- 250g (8oz) fresh peas

BRAISED DUCK WITH PEAS

Preheat the oven to 220°C (425°F), Gas Mark 7.

Put your duck into a heavy casserole lined with the bacon. Arrange the carrot, onion and chestnuts around the duck, then add the bouquet garni and cook on top of the stove over a medium-high heat for 15 minutes with the lid on, turning the duck so that it is browned on all sides. Add the wine and stock, cover the pan again and put into the oven for about 1 hour.

About 15 minutes before the end of the cooking time, melt the butter in a frying pan and cook the chipolatas and the shallots. Keep warm.

When the duck is ready, remove it from the pan and place it on a board. As with roasting, you will know the duck is done when you can stick a skewer into the meat, at a thick point in the leg, and the juices run clear, not bloody. Pour the braising liquid into a smaller pan and reduce to a gravy consistency over a medium heat.

Meanwhile, cook the peas in salted boiling water until tender. Place the duck on a serving dish, then strain the sauce and pour some of it over the duck. Serve with the chipolatas, the peas and the remaining sauce in a jug.

It's all about preparation

METHOD #27

CRISPY DUCK

Crispy duck and pancakes is one of those Chinese delicacies that never fails to impress. Luckily it's easy to make, giving you an extremely tasty dish with all the lovely flavours you thought you could only get in a restaurant. It is also a very sociable meal and great fun to share with family and friends. Once the preparation is done, everyone can sit down together to share the fillings and make their own pancakes. You can add aromatic spices like star anise to the honey mixture for extra flavour.

CRISPY DUCK WITH PANCAKES

Serves 8

1 x 2.5kg (5lb) duck
2 tablespoons honey, mixed with 1 tablespoon water

TO SERVE

Chinese pancakes
shredded spring onions
strips of cucumber
plum sauce

Follow the instructions opposite for making crispy duck. Then serve the duck with Chinese pancakes, accompanied by bowls of spring onions, cucumber and plum sauce.

HOW TO MAKE CRISPY DUCK

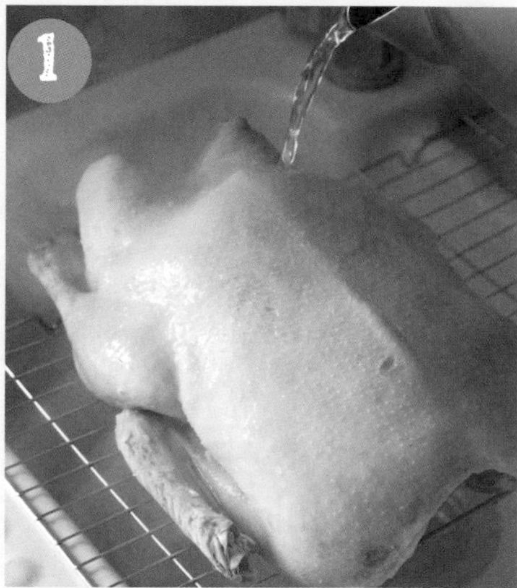

Trim the duck of any excess fat and remove the wing tips and the parson's nose (the fleshy part of the tail). Ensure that the cavity opening is clear. Place the duck on a rack over the sink and slowly pour a full kettle of boiling water over it.

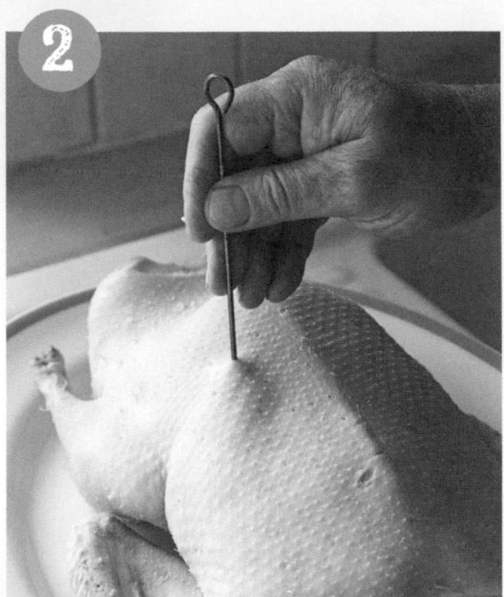

Pat the duck dry and, while it's still warm, prick the skin all over.

Brush it with the honey and water mixture. Put the honey-covered duck to cool and rest in the fridge for at least 5 hours.

Once the duck has rested, preheat the oven to 200°C (400°F), Gas Mark 6. Place the duck on a trivet in a roasting tray and cook without basting for 1½ hours. Check it after 1 hour, and if it is blackening cover it loosely with foil – dark brown is the colour you want.

Once the duck has finished roasting, allow it to rest for at least 15 minutes, then use a fork to shred the meat.

METHOD #28

POULTRY JOINTS

A bird is very versatile and can be treated in lots of different ways to make a staggering variety of meals. More importantly, if you have reared your own bird for the table you will want to get the most from it, so using different joints in different dishes makes a lot of sense. It is worth doing more than one bird at a time: you can freeze the separate joints to give you the ready-prepped ingredients for the recipes that follow.

If you aren't cooking your bird whole then jointing is an essential skill to learn. All fowl have similar anatomy, so once you are confident jointing a chicken, a turkey is just a bigger bird and a partridge smaller. True, duck and goose are slightly different but if you think of them as being 'longer' it all makes sense, though it is not common to split their legs into drumsticks and thighs.

Joints are used in dishes to provide portion-sized servings and they can also make a dish look more sophisticated. For example, it only takes a slight adjustment to this method for you to leave the first wing bone in and that will change the appearance of the breast so that it looks like part of a bird – this is particularly good for small fowl. You may even consider boning out the thighs and drumsticks for ease of eating.

Most people are used to buying a skinless breast fillet, but it wasn't too long ago that chicken breasts were always bought with bone and skin attached. Our jointing method is almost a compromise in that it provides breast fillets, but with skin on – why would you not take every opportunity to have crispy, tasty skin?

THIGHS

If you believe the old adage that 'fat is needed for flavour', thighs are probably the most flavoursome part of a bird. The brown meat of the thigh has a much more complex structure than the white meat of the breast as it is made up of numerous muscles. On a free-range bird that has spent its life scratching around, the leg meat makes up a proportionally larger amount of the total meat and its single bone means it is very simple to handle. A thigh is sufficient meat for a single portion, especially if filleted and cut into smaller pieces, and as thighs are great slow-cooked in flavoursome liquor or even flash-grilled on a barbecue, they can be used in most dishes.

DRUMSTICKS

Drumsticks are the ultimate finger food, and the variety of rubs and marinades you can use on them is endless, so there are flavour combinations to suit all palates.

BREASTS

The breast is the most prized of the joints, with boned, skinless fillets proving the most popular, as lean white meat is thought to be particularly healthy and good for you.

However, there is a lot to be said for a breast cooked with crispy skin, and with a little imagination a finely cut breast can make a whole meal. It doesn't matter how you cook your breast, but it is essential to make sure it is not overcooked or it can become dry and tough. For larger birds it is possible to get two good portions out of a single breast.

WINGS

By cutting a little of the breast meat off along with the wings it can become a real portion rather than just a nibble. There is a whole industry serving spicy, precooked wings, yet very few people think about cooking their own from scratch. As with drumsticks, any rub or marinade can be great.

NOW TRY: MAKING STOCK

When you are jointing the bird, put anything you cut off into a pan with some vegetables (rough chunks of carrot, onion, leek or any winter vegetable are great) to make stock or a broth. Cover the bones, skin and vegetables with water and simmer gently for a couple of hours. When the stock is cool, remove as much fat as you wish and strain the stock through a colander into a bowl. Discard the vegetables in the colander and keep the remaining pieces of chicken for broth.

NOW TRY: MAKING BROTH

After making the stock, pick any remaining pieces of chicken off the bones and add them to some stock with some pearl barley and fresh diced vegetables to make a broth.

HOW TO JOINT A CHICKEN

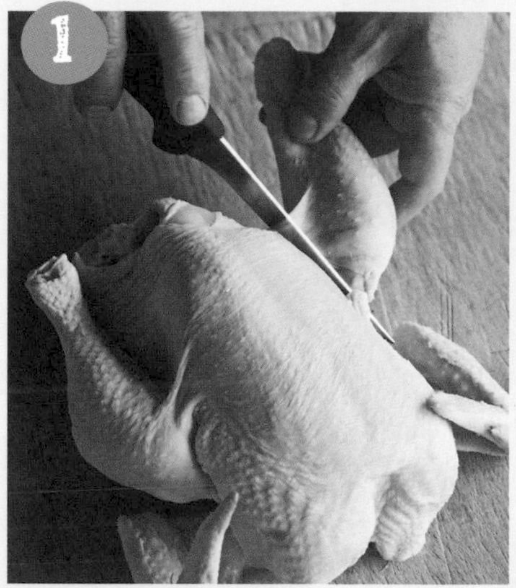

Pull the first leg away from one side of the carcass and cut the loose skin around the leg so you can see the ball and socket.

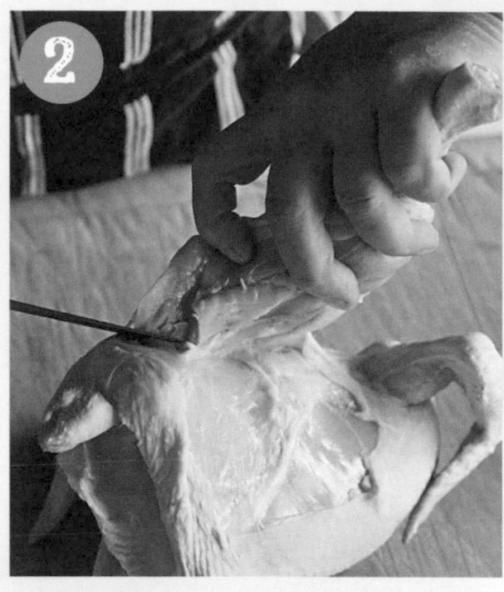

Suspend the carcass from the leg and cut the meat close to the carcass to maximize the amount of meat on the leg.

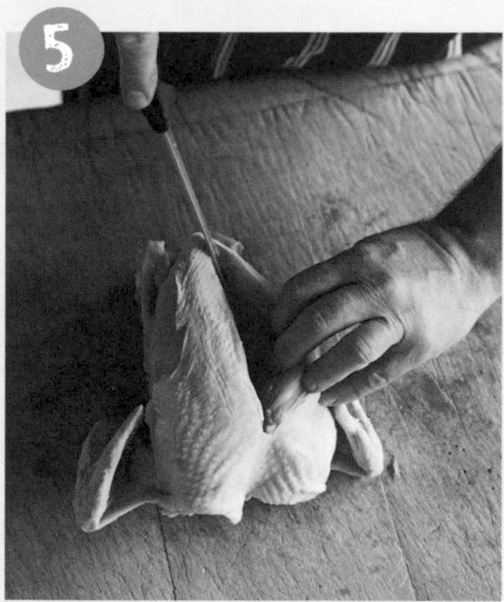

To remove the first breast, slice through the flesh to one side of the breastbone, staying close to the bone, until the breast comes away from the carcass.

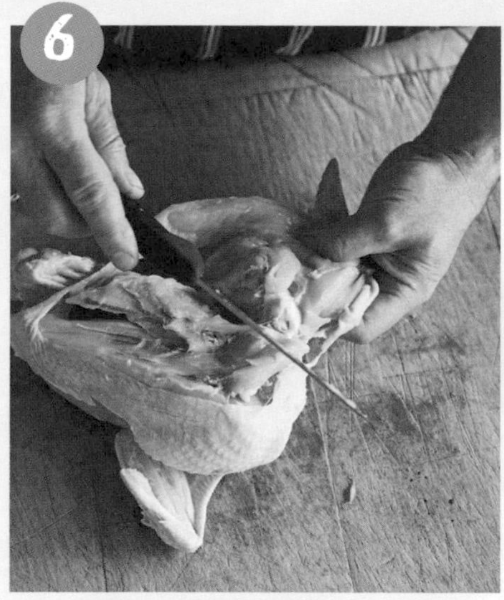

Work forward towards the wing, then, once the breast is free, cut through the wing joint, detaching the wing and breast together. Repeat on the other side.

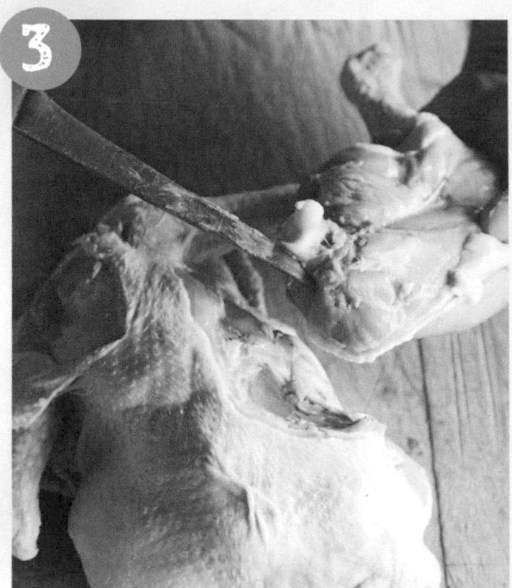

Sever the leg by cutting through the joint with the point of a knife.

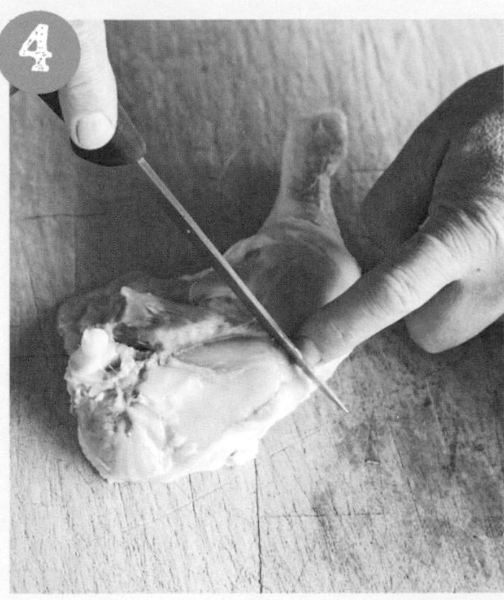

Cut through the join between the thigh and the drumstick, then set the joints aside. Repeat with the second leg, thigh and drumstick.

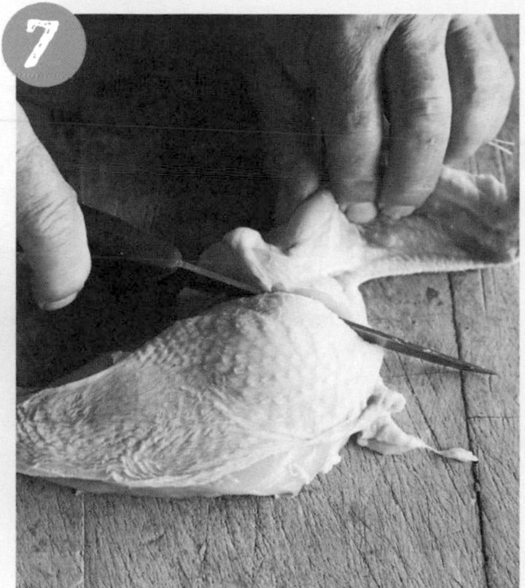

Cut the wing from each breast by slicing through the breast meat.

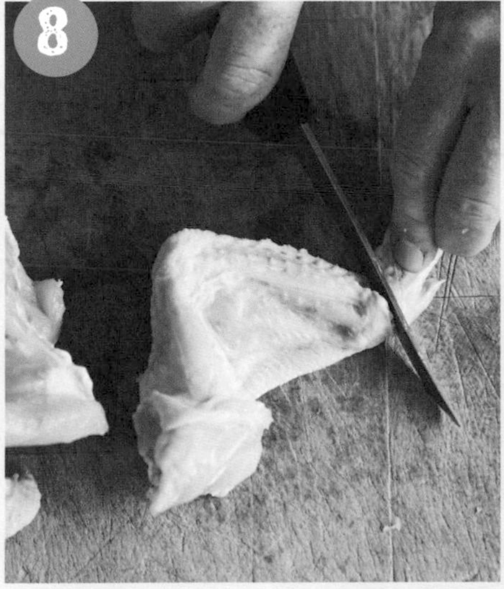

Trim off the tip of the wing by cutting through the joint. The tip has no meat but can be used for stock.

Fried chicken is delicious, and if you want to really enjoy eating drumsticks and chicken wings the answer is to cook them yourself. Made with good, fresh ingredients, this is the perfect finger food.

SERVES 4–6

75g (3oz) plain flour

2 large eggs

50g (2oz) dried breadcrumbs

4 teaspoons paprika

1 teaspoon chilli powder

1 teaspoon freshly ground black pepper

1 teaspoon celery salt

1 teaspoon dried oregano

1 teaspoon chopped fresh sage

1 garlic clove, very finely chopped

4 sweetcorn cobs

800g (1lb 10oz) chicken – drumsticks, wings and thighs

vegetable oil, for deep-frying

FOR THE BARBECUE BEANS

100ml (3½ fl oz) orange juice

75ml (3fl oz) soy sauce

2 tablespoons honey

a splash of Worcestershire sauce

200g (7oz) tinned haricot beans, drained

SOUTH-WESTERN FRIED CHICKEN

Put the flour on to a plate and beat the eggs in a bowl. Put the breadcrumbs into a separate bowl with all the spices, herbs and garlic. Set aside until you are ready to cook the chicken.

Over a high heat, griddle the sweetcorn on a dry ridged pan for 10 minutes, turning regularly.

Meanwhile, make the barbecue beans. Put the orange juice, soy sauce, honey and Worcestershire sauce into a pan with the beans and heat for about 5 minutes. Remember to keep turning the sweetcorn. Keep the beans and sweetcorn warm.

While the sweetcorn is cooking, heat some vegetable oil to 180°C (350°F) in a deep-fryer or large pan. Dip each piece of chicken into the flour first, dusting off any excess, then into the beaten egg. Once coated in egg, quickly transfer the chicken pieces to the aromatic breadcrumbs and roll them around until evenly covered. Deep-fry just 2 or 3 pieces at a time, to avoid lowering the temperature of the oil. Cook for 5 minutes (more for fat thighs), or until golden outside and cooked through inside, then drain on kitchen paper.

Serve the fried chicken with the sweetcorn and beans.

Roasting, slow-cooking and confit-style cooking are delicious methods, but they often take a long time. This recipe is quick and easy, in contrast, and provides a special barbecue grill in minutes -- teriyaki-style sticky duck!

SERVES 2

FOR THE DUCK

150ml (5fl oz) mirin (sweet rice wine), or sake plus 1 teaspoon of sugar

200ml (7fl oz) soy sauce

25ml (1fl oz) rice vinegar

60g (2¼oz) brown sugar

1 tablespoon sesame oil

4 garlic cloves, finely chopped

1 teaspoon grated fresh ginger

a pinch of red pepper flakes or black pepper

4 duck legs

4 duck drumsticks

FOR THE SALSA

1 ripe mango

1 tablespoon chopped fresh coriander

½ a fresh red chilli, finely diced

juice of 1 lime

STICKY DUCK WITH MANGO SALSA

Put the mirin into a pan and bring to the boil, then reduce the heat to medium and leave to simmer for 10 minutes. Add the soy sauce, vinegar, sugar, sesame oil, garlic, ginger and pepper flakes and heat through, then pour into a large bowl and set aside to cool.

At least 20 minutes before you are going to cook the duck, put the pieces into the bowl of sauce and leave to marinate until you are ready to cook them. Heat a grill or griddle until hot and cook the duck, brushing it with the sauce, turning the pieces once and adding more sauce if you want to make them extra sticky.

While the duck is cooking, make the salsa. Simply cut the mango into small cubes and put them into a bowl, then add the chopped coriander, chilli and lime juice. Serve with the duck.

Cook on a griddle – and add more sauce for extra stickiness

Caesar dressing is a classic – this is our version. We are always looking for extra ways to use our hens' eggs and this salad provides the perfect opportunity. The key with this dish is not to overcook the chicken, and here it is cooked quickly in a hot pan so that it stays moist.

SERVES 4

- 100g (3½oz) Parmesan cheese
- 3 garlic cloves, chopped
- 6 anchovy fillets, chopped
- 1 teaspoon Worcestershire sauce
- juice and zest of 1 lemon
- 1 egg
- 4 tablespoons olive oil, plus extra for frying
- 2 thick slices of bread
- 2 teaspoons fresh lemon thyme leaves
- salt and freshly ground black pepper
- 450g (14½ oz) chicken breasts, cut into chunky strips
- 2 rashers of smoked streaky bacon, cut into small pieces
- 2–3 little gem lettuces

CHICKEN CAESAR SALAD

Grate half the Parmesan and put it into a blender with 1 of the garlic cloves, or whisk in a bowl. Add the anchovy fillets, Worcestershire sauce, lemon juice and egg. Blend until smooth, then gradually add up to half the olive oil, stirring until the dressing is emulsified. Taste and set aside.

To make the croutons, cut the bread into 5mm (¼ inch) cubes and put them into a bowl. Add half the lemon thyme. Chop the remaining 2 garlic cloves and add to the bowl, reserving a little of the chopped garlic for cooking the chicken, if you like. Season with salt and pepper. Add 2 more tablespoons of oil and toss everything together well. Fry the croutons in a hot frying pan until golden brown all over, then leave to drain on kitchen paper.

Put the chicken into a bowl and season with salt, pepper, the lemon zest, the reserved garlic, if using, and the rest of the thyme. Heat some oil in a frying pan, add the chicken and bacon and cook quickly until the chicken is tender, removing the bacon when it gets crisp.

Meanwhile, wash and dry the lettuce leaves and arrange them in a serving dish. Top with the croutons and the warm chicken and bacon. Shave the remaining Parmesan into thin pieces and scatter over the salad, then drizzle over the dressing.

This soup is full of nourishment and it's a real winter warmer. You can adjust the ratio of ingredients, if you like, and the soup can be cooked in a pressure cooker or a conventional large pan. We don't mess around with this broth – it all goes straight into one pot from the beginning.

SERVES 4

- 8 chicken thighs
- 2 onions, halved and thinly sliced
- 8 garlic cloves, sliced
- 3 fresh chillies, chopped (seeds and all)
- 75g (3oz) fresh ginger, peeled and chopped into fine matchsticks
- 4 tablespoons dried goji berries or cranberries
- 1 litre (1¾ pints) chicken stock
- salt and freshly ground black pepper

SPICY CHICKEN BROTH

Trim any excess fat off the chicken thighs, then put them into a large pan or a pressure cooker with the rest of the ingredients. Bring to the boil, skimming off any scum that rises to the surface.

If you are using a pressure cooker, put the lid on and cook for 30 minutes. If you are using a conventional pan, simmer the broth for at least 2 hours.

Allow the soup to cool, then take the chicken thighs out and remove the skin and bones (you can also serve it bones and all, if you prefer). Return the chicken meat to the broth, season with salt and pepper, then reheat and serve.

METHOD #29

CONFIT DUCK

This slow-cooking method involves curing the duck in salt and then cooking it in lard until the brown meat is so soft you could eat it with a spoon. It produces very tender meat that can be stored for 3–5 months. Traditionally, confit duck is cooked in its own fat, but we use lard as it solidifies better for storage. The confit duck can be stored in a jar in the fridge as long as it is covered completely with the cooking fat. To serve, bring to room temperature and wipe off any excess fat. You can also use the same method to confit goose and turkey. The end result is very rich, so serve it with something a little sharp.

HOW TO CONFIT DUCK LEGS

1

First prepare your cure by mixing the garlic, peppercorns, bay leaves, fresh thyme and salt in a bowl. Rub this mixture into the duck legs, then put them into a dish and leave them overnight in the fridge.

2

Preheat the oven to 120°C (250°F), Gas Mark ½. Put the lard into a pan and heat gently until melted. Rinse the duck legs and dry them. Put them into an ovenproof dish, cover with the melted lard and cook in the oven for 3 hours. Remove from the oven and allow to cool, then transfer to a jar and store in the fridge.

CONFIT DUCK WITH CARAMELIZED ORANGE & FENNEL SALAD

Serves 2

FOR THE CONFIT

2 garlic cloves, chopped

10 peppercorns

2 bay leaves

sprigs of fresh thyme

5 tablespoons salt

2 duck legs (about 300g/10½ oz)

500g (1lb) lard

FOR THE CARROT PURÉE

2 carrots, peeled and chopped

FOR THE SALAD

50g (2oz) butter

½ teaspoon sugar

2 oranges, sliced with skin on

2 fennel bulbs, sliced

Make the confit following the instructions opposite.

For the purée, boil the carrots until tender, drain, then place in a food processor and blend until smooth.

To make the salad, heat the butter and sugar in a frying pan, then add the orange slices and cook until caramelized on one side. Turn them over and do the same on the other side, then leave to cool, saving the pan juices. Meanwhile, chargrill the fennel.

Arrange slices of fennel and orange around some carrot purée. Top with a leg of duck and drizzle with the pan juices.

NOW TRY: CONFIT GOOSE

The principles of making a confit are transferable to other fowl and you can vary the herbs and spices to suit your personal preferences.

For goose try adding star anise, fresh ginger and garlic to the confit, or you could try a mixture of sage, bay and peppercorns. You may need to increase the cooking time a little if it is a large goose.

NOW TRY: CONFIT TURKEY

A turkey drumstick makes an impressive confit, though you will need to draw out the sinews before you start so that the final dish can be served easily. It is tasty with a combination of cinnamon, cardamom, coriander seeds and cumin.

METHOD #30

BONED BIRDS

Boning is a fiddly job and it takes practice to do it quickly, but the benefits are well worth the effort. Once you've boned a bird, you will have a casing that can be stuffed, roasted and then carved easily into neat slices that contain white meat, dark meat and stuffing. It's simplest to learn to bone with a chicken, as we've shown here. Once you've mastered that you can bone other game birds and domestic fowl the same way

HOW TO BONE A CHICKEN

1

Using a boning knife, begin by removing the wing tips by putting the tip of the knife into the joint and pressing down firmly. Set them aside in a pan, together with the other bones as you come to them, ready to make stock.

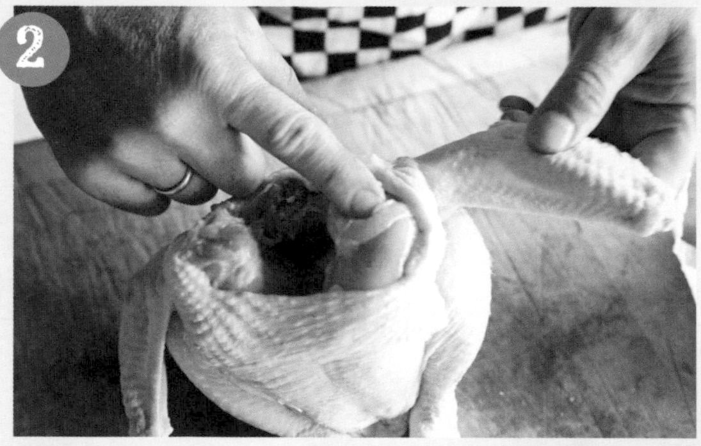

2

Slip your fingers under the skin at the neck hole, and work the skin back until you expose one of the shoulder joints. Using the tip of your knife, cut through the shoulder joint to separate the wing bones from the carcass. Repeat the process with the other wing.

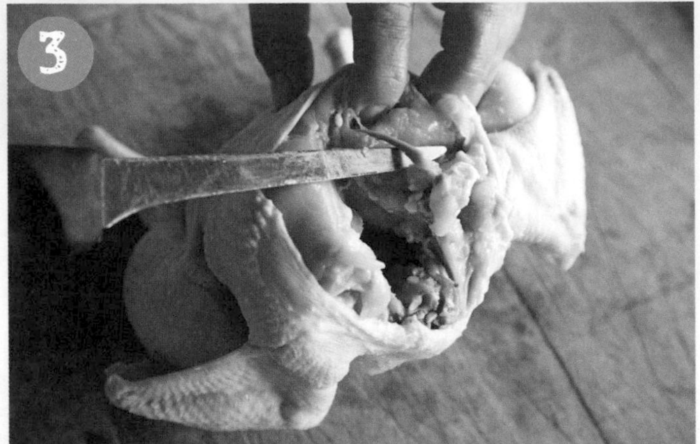

Pull the skin a little further back over the breast to expose the wishbone.

Use the tip of the knife to scrape away the flesh around the wishbone. Remove the wishbone.

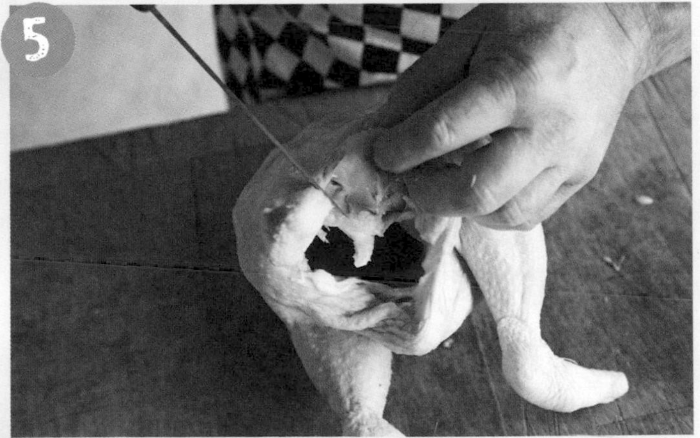

Place the bird breast side down, remove the parson's nose (the fleshy part of the tail) and cut the skin from where the parson's nose was, all the way down to the hip joint of one of the legs.

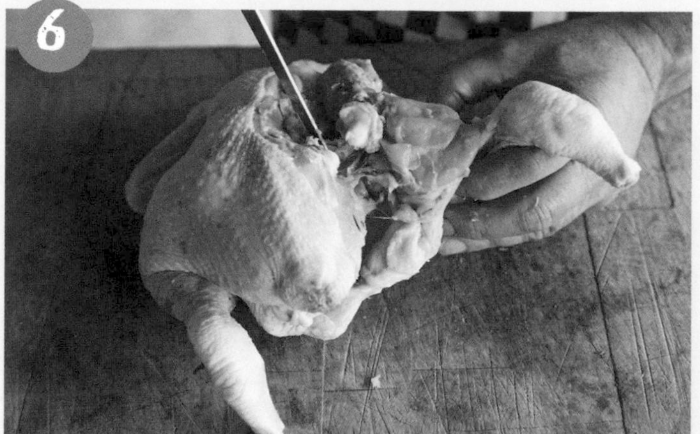

With the point of the knife, separate the hip joint from the carcass, leaving the bone in place.

You are now ready to trim the meat from the carcass, so with the breast side down, cut straight along the backbone in long gentle slices.

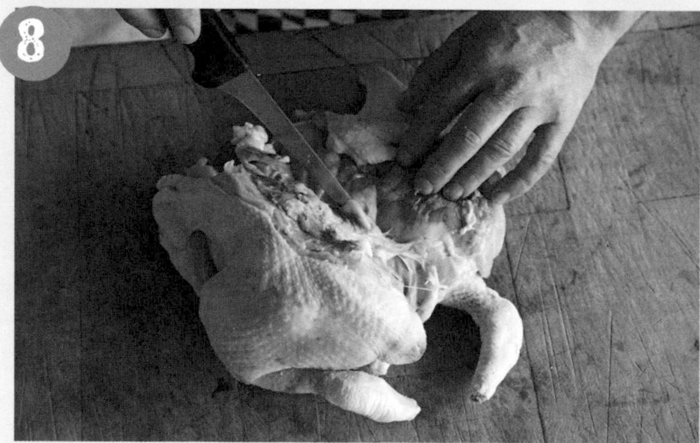

Keep the blade near the ribs, pushing the chicken flesh back as you go; the effect will be rather like turning the chicken inside out.

As you are cutting down along the carcass be careful during this phase to keep the blade of the knife pointed towards the carcass, and be particularly careful when you get to the top of the breastbone lest you inadvertently puncture the skin.

Once you have freed all the breast from one side of the carcass, turn the bird around and repeat steps 6–9 on the other side of the bird.

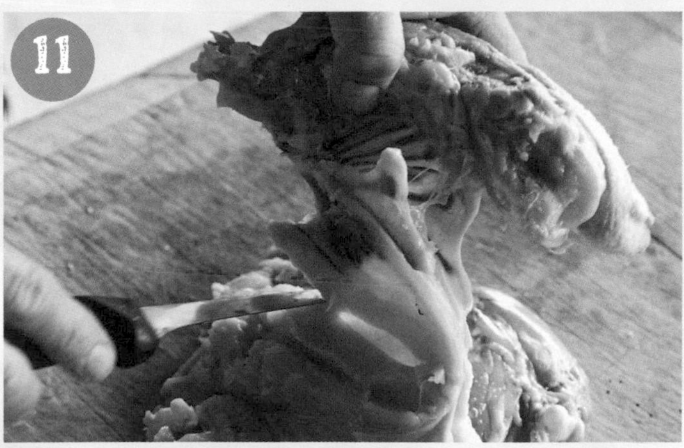

Then hold the carcass and allow the meat to hang beneath it. Cut horizontally to separate the carcass from the meat – don't cut the skin!

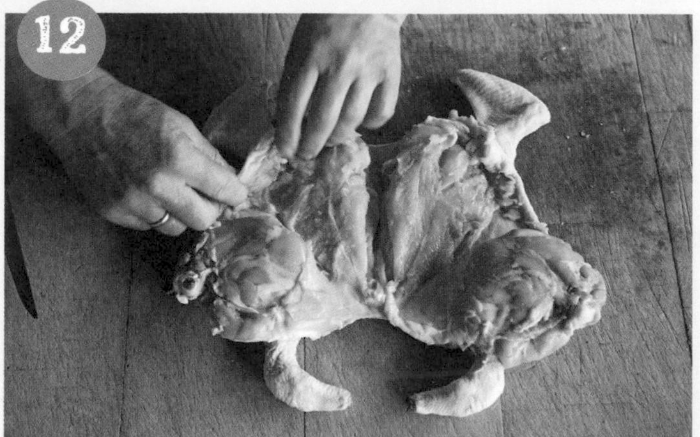

Now it's time for the legs. Grasp the thigh bone at the hip end and start to separate the meat from the thigh.

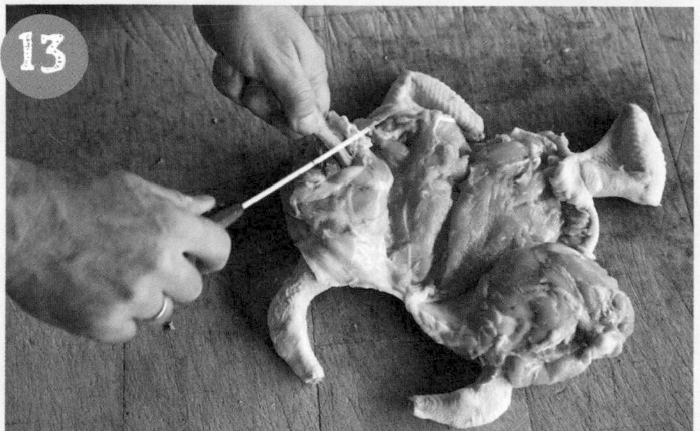

Using a knife, scrape the meat off the thigh bone.

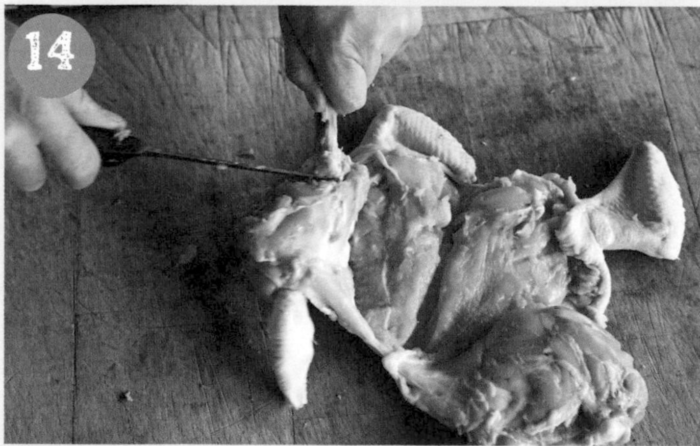

At this stage you can cut off the thigh bone, leaving the drumstick in, or continue stripping back the meat and taking out both the bones.

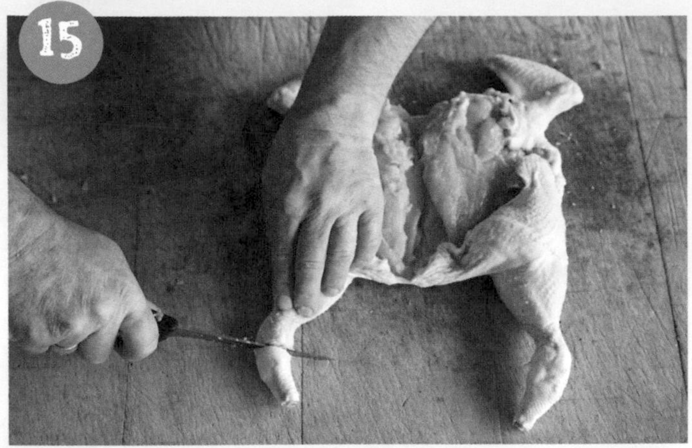

Finally, cut off the end of the leg. Take the boned meat and skin and re-form it back into the shape of the bird. It is now ready to stuff.

STUFFING A BONED BIRD

To stuff the boned bird it is easiest to put it breast side down, place your stuffing on the breast meat and push the stuffing into the thigh cavities. Then weave a skewer through the skin to join the edges together or (an easier option) pull together the edges of the skin and sew them together. If you leave the bones in the wings and drumsticks, the final dish when presented at the table will look very like a normal roasted bird.

the bones - ready for the stock pan

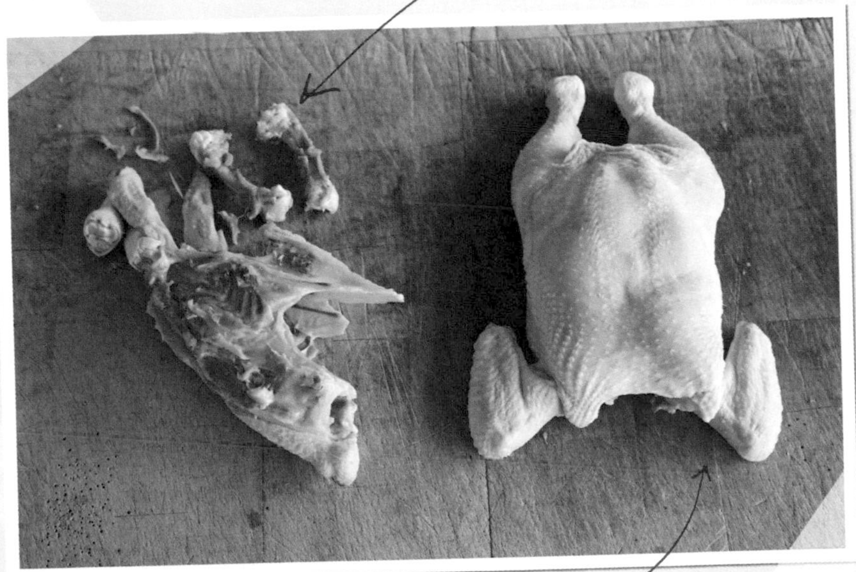

the stuffed bird

Southern-fried chicken is a popular treat and we have created our own take on this familiar dish. The crispy, breadcrumbed chicken goes well with coleslaw. You can easily change the herbs and spices in the breadcrumbs to create different flavours – try lemon zest and lots of cracked black pepper, jerk spices or a mixture of cumin, coriander, chilli and turmeric.

SERVES 4

FOR THE CHICKEN

4 skinless, boneless chicken breasts

2 tablespoons plain flour

1 egg, beaten

200g (7oz) breadcrumbs (day-old dry crumbs work best)

salt and freshly ground black pepper

zest of ½ a lemon

a little smoked paprika

a pinch of cayenne pepper

vegetable oil for shallow frying

FOR THE COLESLAW

2 apples

1 tablespoon lemon juice

¼ of a white cabbage, thinly shredded

4 carrots, shredded

1 tablespoon mayonnaise

FRIED CHICKEN WITH 'SLAW

First make your coleslaw. Peel the apples and grate them into a bowl. Stir in the lemon juice and season with salt and pepper. Add the cabbage, carrots and mayonnaise and mix until everything is coated.

Cover and leave in the fridge until you need it.

Butterfly the chicken fillets (see page 164), then lay them one at a time between 2 pieces of clingfilm on a board and bash them until they are about 5mm (¼ inch) thick. Prepare a breading station of bowls containing flour, beaten egg and breadcrumbs. Flavour the breadcrumbs with salt and pepper, lemon zest, paprika and cayenne. Be generous with the salt. Dip each piece of chicken into the flour first, then into the beaten egg and finally coat them with the breadcrumbs, patting them on so that they are nice and thick.

Heat some vegetable oil in a pan, but don't let it get too hot – you want the chicken to cook through rather than burning on the outside. Cook the chicken for 3–4 minutes on each side, until golden. Leave to stand for a couple of minutes, then serve with the coleslaw.

Barbecued chicken can take many different forms, and this version puts an Asian twist on traditional chicken skewers – the charcoal taste really works well with the peanut sauce.

SERVES 4

450g (14½ oz) chicken breasts

FOR THE SATAY SAUCE

2 tablespoons peanut oil

½ a white onion, chopped

1 fresh red chilli, seeded and finely chopped

1 garlic clove, finely chopped

100g (3½oz) smooth peanut butter

juice of 2 limes

2 tablespoons soy sauce

2 teaspoons sugar

FOR THE MARINADE

1 tablespoon soy sauce

1 teaspoon freshly ground black pepper

1 teaspoon lime zest

1 tablespoon crushed peanuts

1 tablespoon sesame oil

TO SERVE

thin slices of spring onion and cucumber

lime wedges

BARBECUED SATAY SKEWERS

To make the satay sauce, heat the peanut oil in a wok or frying pan and add the onion. Cook until golden, then add the chilli and garlic. Add the peanut butter, then take off the heat and stir in the lime juice, soy sauce and sugar. Set aside to reheat later.

Soak 8–12 bamboo skewers in water for at least 30 minutes. Cut the chicken breasts across the grain into thin strips. Mix the marinade ingredients together in a shallow dish, add the chicken strips, then put into the fridge and leave to marinate for 1–2 hours. Drain away the marinade and thread the skewers through the chicken strips in simple S-shapes.

Get your barbecue good and hot, then grill the chicken skewers for 5 minutes or until cooked through. If you don't have a barbecue, you can flash-fry the chicken skewers in a frying pan or cook them in the oven at 180°C (350°F), Gas Mark 4, for 10 minutes. When the skewers are nearly ready, reheat the satay sauce over a low heat.

Serve with thin slices of spring onion and cucumber, lime wedges and your homemade satay dipping sauce.

Outdoor cooking on our Kotlich

The lemon pepper we use in this recipe adds fantastic flavour to an otherwise simple chicken dish. Making a powder for flavouring poultry is incredibly quick and easy and avoids all the chemicals and preservatives used in so many shop-bought seasoning mixes.

SERVES 4

450g (14½ oz) chicken breasts

2 tablespoons plain flour

1 egg

50g (2oz) breadcrumbs

vegetable oil, for deep-frying

lemon wedges, to serve

mayonnaise (optional)

FOR THE LEMON PEPPER

zest of 4 lemons

1 teaspoon fresh lemon thyme leaves, finely chopped

1 teaspoon salt

1 teaspoon freshly ground black pepper

LEMON PEPPER CHICKEN NUGGETS

To make the lemon pepper, spread the lemon zest thinly and evenly on a microwave dish. Microwave for 6–8 minutes on a medium power setting. The lemon will be ready when it feels dry and crispy but before it starts to turn golden brown or looks burnt. (If you don't have a microwave, spread the lemon zest thinly on a baking tray and put in an oven at 50°C/120°F for 4–5 hours until dry.) Leave to cool, then blend in a coffee grinder until fine. Mix with the lemon thyme leaves, salt and pepper and store in a sealed jar in a cool, dry, dark place, where it will keep for 3–4 weeks.

Cut the chicken breasts into bite-sized pieces. Put the flour onto a plate and beat the egg in a bowl. Mix the breadcrumbs with 2 tablespoons of the lemon pepper in a shallow dish. Dip the chicken pieces into the flour first, then into the beaten egg, and finally into the flavoured breadcrumbs.

Heat some vegetable oil to 180°C (350°F) in a deep-fryer or a large pan and deep-fry the chicken in batches for 4–5 minutes, until golden outside and cooked through inside. Alternatively, you can shallow-fry it in a couple of tablespoons of oil for 5–10 minutes. Drain on kitchen paper and serve with lemon wedges.

For a quick and easy dip, mix 1 teaspoon of the lemon pepper with 2 tablespoons of mayonnaise and serve with the chicken.

Here we use well-loved turkey accompaniments, such as chestnuts and apple, to make a burger with real depth of flavour – and the fresh, homemade cranberry sauce gives it zing. This recipe works equally well with leftover cooked turkey.

SERVES 4

FOR THE CRANBERRY RELISH

250g (8oz) caster sugar

100ml (3½ fl oz) water

250g (8oz) cranberries

zest and juice of 1 large orange

1 stick of cinnamon

1 clove

FOR THE BURGERS

625g (1¼ lb) turkey meat

1 large onion, finely chopped

1 tablespoon chopped fresh parsley

100g (3½ oz) peeled chestnuts

1 apple, finely diced

1 egg, beaten

20g (¾ oz) fresh breadcrumbs

zest of 1 lemon

salt and freshly ground black pepper

vegetable oil, for frying

TO SERVE

seeded bread rolls, split in half

slices of grilled bacon

a few salad leaves

THE ULTIMATE TURKEY BURGER

To make the relish, put the sugar and water into a thick-bottomed pan and heat until the sugar has dissolved. Add the cranberries and the orange zest and juice. Bring to the boil, then simmer over a medium heat for 5 minutes. Add the cinnamon and the clove, turn up the heat and cook for a further 10 minutes. Take off the heat and allow to cool slightly before blending in a food processor to make a rough relish. Set aside until completely cool, then transfer to a bowl and refrigerate.

Roughly chop the turkey meat and blitz it quickly in a blender or food processor. Put it into a mixing bowl and add all the other ingredients except the oil. Mix well, then shape into large balls and flatten them slightly. Heat a couple of tablespoons of vegetable oil in a frying pan and fry for 5 minutes on each side, until cooked through.

Serve your turkey burger in a bread roll, with a slice or two of grilled bacon, some salad leaves and a generous spoonful of cranberry relish.

Burgers with a difference

METHOD #31

ROULADE

A roulade is a surprisingly easy way to serve poultry. All you need is skinless, boneless chicken breasts and the filling of your choice. The meat will be infused with flavour and moist from poaching. Try fillings such as garlic and lemon, pesto, spinach and ricotta, or walnut and sage.

A FILLING FOR CHICKEN ROULADE

For 4 skinless, boneless chicken breasts (about 450g / 14½oz)

- 150g (5oz) sundried tomatoes, chopped
- 4 garlic cloves, chopped
- juice of 2 lemons
- olive oil
- 2 teaspoons paprika
- 2 teaspoons fresh tarragon leaves

Mix together all the ingredients except for the tarragon in a pestle and mortar.

HOW TO MAKE A ROULADE

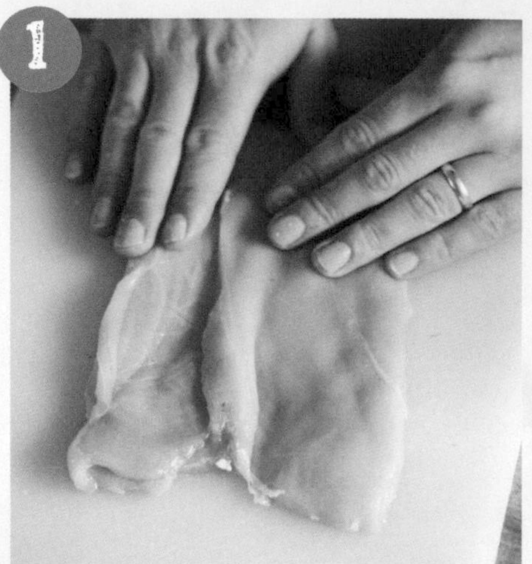

1

Using a sharp filleting knife, cut the breast along its length from one side. Stop just before you reach the other side so that you can fold it out flat.

4

Roll the meat into a cylinder, gathering the clingfilm as you go.

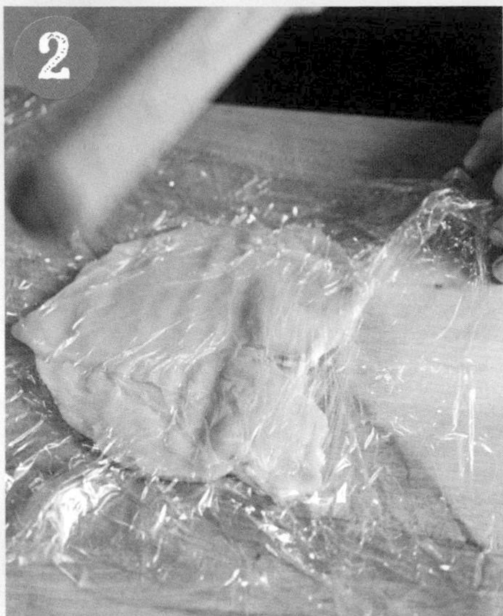

Place the meat between 2 pieces of clingfilm and bash it gently with a rolling pin until it is about 5mm (¼ inch) thick all over. Don't be too heavy-handed.

Peel off the top piece of clingfilm, then generously and evenly spread your filling over the meat, like spreading a paste. Sprinkle on the tarragon leaves.

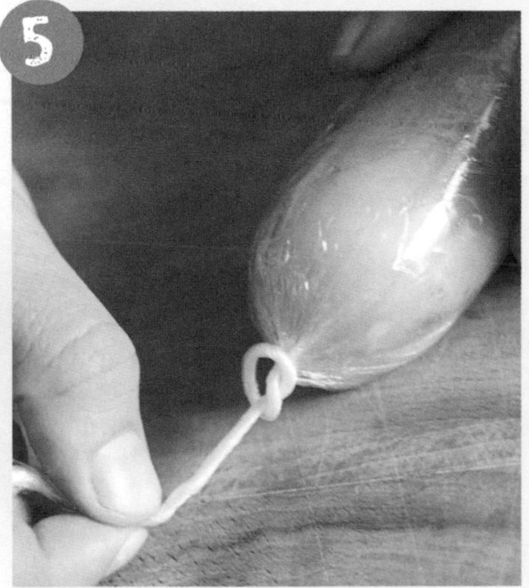

When you are happy with the feel of your roulade – it should be firm and cylindrical – tie off the ends with a simple granny knot.

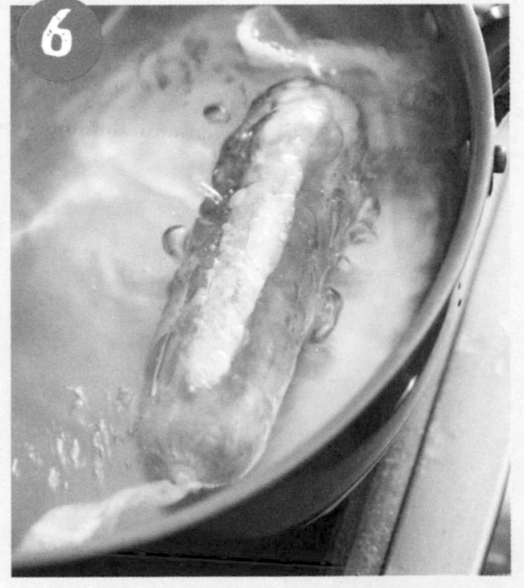

Bring a large pan of water to the boil, then add the roulade (or roulades) and poach for 15–20 minutes. When ready, remove the clingfilm, place on a carving board, slice and serve.

Chicken livers are plentiful, cheap and very tasty. This simple pâté goes a very long way and is great with crispy toast or brioche. For a change of flavours you could add a little lemon thyme or use port instead of brandy.

SERVES 6

- 200g (7oz) chicken livers
- 150g (5oz) lean rindless bacon
- 100g (3½oz) softened butter
- 3 egg whites
- 2 tablespoons brandy
- salt and white pepper

CHICKEN LIVER PÂTÉ

Preheat the oven to 160°C (325°F), Gas Mark 3.

Remove any connective tissues and discoloured flesh from the chicken livers, then cut the bacon into small pieces. Place all the ingredients in a food processor and whizz until smooth.

Put the mixture into a terrine or other ovenproof dish, then stand it in a roasting tray and pour in water to come halfway up the sides of the terrine (making a bain-marie). Cook in the oven for 45 minutes, then remove and allow to cool.

Chill thoroughly before serving.

whizz it all together

If you haven't tried goose livers before, give them a chance and cook them with gusto! To cut through their rich flavour, we paired the livers with a sharp cider and sweet apple sauce for this recipe.

SERVES 2–4

300g (10oz) goose livers
50g (2oz) plain flour
salt and freshly ground black pepper
2 tablespoons vegetable oil
2 teaspoons chopped fresh parsley
fresh chive flowers (optional)

FOR THE APPLE SAUCE

2 large sweet apples
20g (¾ oz) butter
2 tablespoons sugar
1 bottle of cider
1 tablespoon wholegrain mustard

GOOSE LIVERS WITH CIDER

First make the apple sauce. Peel your apples, cut them into small cubes, then put them into a pan with the butter and sugar over a medium heat until they start to soften and release their juice. Add 300ml (½ pint) of cider and continue to cook until it has reduced by half. Stir in the mustard. Set the sauce aside and keep warm. (Keep the rest of the cider to enjoy with the goose livers.)

Slice the goose livers thinly. Dip the slices into seasoned flour and shake off the excess. Heat the vegetable oil in a frying pan and cook the livers for 1–2 minutes on each side, or until the edges start to turn golden brown.

Serve warm, with the apple sauce, and garnish with the parsley and some chive flowers, if you have them.

be careful not to overcook the livers

INDEX

ACKNOWLEDGEMENTS

Publisher: Stephanie Jackson
Managing Editor: Clare Churly
Copy-editor: Annie Lee
Creative Director: Jonathan Christie
Designer: Jaz Bahra
Illustrators: Charlotte Strawbridge, James Strawbridge
Photographer: Nick Pope
Stylist: Alison Clarkson
Kitchen Dogsbody: Jim Tomson
Production Controller: David Hearn

Picture credits

All photographs © **Nick Pope** with the exception of the following: **Alamy** Arndt Sven-Erik/Arterra Piture Library 70; E Westmacott 73; Gary K Smith 69; Helmut Meyer zur Capellen/imagebroker 49a; Juniors Bildarchiv 29; Karen Appleyard 52b; Lyndon Beddoe 47; Photolocate 71; Richard Mittleman/Gon2Foto 84. **Corbis** Lars Langemeier/AB 31; Robin Loznak/ZUMA Press 87b; Steve Maslowski/Visuals Unlimited 50. **FLPA** Cyril Ruoso/Minden Pictures 72; ImageBroker 77br; Patricio Robles Gil/Minden Pictures 83l. **Flyte so Fancy Ltd** www.flytesofancy.co.uk 85. **Fotolia** eag1e 51a. **Getty Images** Photolibrary/Anthony Blake 89b. **Photoshot** ErnieJanes/NHPA 48. **Science Photo Library** Robert Llewellyn/AgstockUSA 87a. **Strawbridge Family Archive** 19al, ar & br, 24r, 25, 41ar & bl, 42ar & br, 46b, 51b, 52a, 67, 77al, 81, 83r, 89a. **Thinkstock** AbleStock.com 42 al; Hemera 13a, 26, 27, 41 al & br, 42, 46a, 61ar & bl, 62 all, 68, 77ar & bl, 78 all, 82.

Backgrounds: Fotolia/Monica Butnaru; iStockphoto/Thinkstock

Illustrations: **Charlotte Strawbridge** 16, 38, 58, 74, 90. **James Strawbridge** 23, 31, 65, 81.